just

THE JOB

Land & the environment

FOR SALE

PRICE 20p

just **THE JOB**

Land & the environment

Lifetime Careers
WILTSHIRE

Hodder & Stoughton

A MEMBER OF THE HODDER HEADLINE GROUP

Just the Job! draws directly on the CLIPS careers information database developed and maintained by Lifetime Careers Wiltshire and used by almost every careers service in the UK. The database is revised annually using a rigorous update schedule and incorporates material collated through desk/telephone research and information provided by all the professional bodies, institutions and training bodies with responsibility for course accreditation and promotion of each career area.

ISBN 0 340 68791 6
First published 1997

Impression number	10	9	8	7	6	5	4	3	2	1
Year		2002	2001	2000	1999	1998	1997			

Printed in Great Britain for Hodder & Stoughton Educational, the educational publishing division of Hodder Headline Plc, 338 Euston Road, London NW1 3BH, by Cox & Wyman Ltd, Reading, Berkshire.

CONTENTS

Introduction   9

Careers using geography   10
Cartography. Surveying. Planning. Landscape architecture. Environmental and ecological work. Meteorology. Oceanography. Teaching. Library/information work. Transport and distribution. Marketing and logistics. Travel and tourism. Work abroad.

Caring for the environment   15

Meteorology   20

Cartography   24
Surveying and photogrammetry. Editorial work. Cartographic drafting.

Land surveying   31

Rural practice surveying   34
Land agency. Rural estate management. Agricultural surveying. Agricultural management. Forest management. Auctioneering, valuation and claims. Technical surveying.

Building surveying   38
Building surveyor. Technical surveyor. Civil engineering surveyor.

Quantity surveying   42

Mining surveying   46

Mining & mining engineering   49

Quarrying   53
Machine operating. Engineering and maintenance. Technician, supervisory and management posts.

Estate agency   57

Valuation & auctioneering   62

Landscape architecture, management & science   68
Landscape architect. Landscape scientists and managers.
Technician and craft-level work.

Town & country planning   74
Local authority work. Planning support staff.

Wastes management   83

The water industry   87

Hydrographic surveying   90

Oceanography   92

The offshore oil & gas industry   95

Geology & geophysics   100

For further information   104

JUST THE JOB!

The *Just the Job!* series ranges over the entire spectrum of occupations and is intended to generate job ideas and stretch horizons of interest and possibility, allowing you to explore families of jobs for which you might have appropriate ability and aptitude. Each *Just the Job!* book looks in detail at a popular area or type of work, covering:

- ways into work;
- essential qualifications;
- educational and training options;
- working conditions;
- progression routes;
- potential career portfolios.

The information given in *Just the Job!* books is detailed and carefully researched. Obvious bias is excluded to give an even-handed picture of the opportunities available, and course details and entry requirements are positively checked in an annual update cycle by a team of careers information specialists. The text is written in approachable, plain English, with a minimum of technical terms.

In Britain today, there is no longer the expectation of a career for life, but support has increased for life-long learning and the acquisition of skills which will help young and old to make sideways career moves – perhaps several times during a working life – as well as moving into work carrying higher levels of responsibility and reward. *Just the Job!* invites you to select an appropriate direction for your *own* career progression.

Educational and vocational qualifications

A level – Advanced level of the General Certificate of Education

AS level – Advanced Supplementary level of the General Certificate of Education (equivalent to half an A level)

BTEC – Business and Technology Education Council: awards qualifications such as BTEC First, BTEC National Certificate/Diploma, etc

GCSE – General Certificate of Secondary Education

GNVQ/GSVQs – General National Vocational Qualification/General Scottish Vocational Qualification: awarded at Foundation, Intermediate and Advanced levels by BTEC, City & Guilds, Royal Society of Arts and the Scottish Qualifications Authority (SQA)

HND/C – BTEC Higher National Diploma/Certificate

International Baccalaureate – recognised by all UK universities as equivalent to a minimum of two A levels

NVQ/SVQs – National/Scottish Vocational Qualifications

SCE – Scottish Certificate of Education, at **Standard** Grade (equate directly with GCSEs: grades 1–3 in SCEs at Standard Grade are equivalent to GCSE grades A–C) and **Higher** Grade (equate with the academic level attained after one year of a two-year A level course: three to five Higher Grades are broadly equivalent to two to four A levels at grades A–E)

Vocational work-based credits	NVQ/SVQ level 1	NVQ/SVQ level 2	NVQ/SVQ level 3	NVQ/SVQ level 4
Vocational qualifications: *a mix of theory and practice*	Foundation GNVQ/GSVQ; BTEC First	Intermediate GNVQ/GSVQ	Advanced GNVQ/GSVQ; BTEC National Diploma/Certificate	BTEC Higher National Diploma/Certificate
Educational qualifications	GCSE/SCE Standard Grade pass grades	GCSE grades A–C; SCE Standard Grade levels 1–3	Two A levels; four Scottish Highers; Baccalaureate	University degree

opportunities – for both graduates and non-graduates – are only summarised here. Other sections go into more detail.

Cartography

Map-making employs only a few people, though there is fairly steady recruitment by both the Ordnance Survey and private cartographic firms. Much surveying work is now done by aerial photography, and computers are increasingly used to draw maps. Entrants are usually recruited either with GCSEs at grade C or a degree (specialist degrees, and options within degrees, are available).

Surveying

Site surveyors are involved in producing plans of sites and studying the nature of the ground, its drainage, etc. **Rural practice surveyors** are concerned with land use and methods of cultivation. **Hydrographic surveying** is another possibility. Geography GCSE at grade C, or A level, is useful though not essential. Mathematics and English GCSE at grade C, or equivalent qualifications, are necessary for all types of surveying, and most surveyors are graduates.

Planning

Planners are concerned with overall land use and detailed studies and plans in both town and country. A geography GCSE at grade C or A level, or a degree, is a useful preparation (specialist degrees are also offered).

Landscape architecture

This is concerned with the design and improvement of landscapes in town and country. There are degree/diploma courses, and an interest in geography (especially physical geography and cartography) is useful. Artistic/design abilities are essential.

Environmental and ecological work

There are posts in local authorities, water companies, countryside conservation organisations and national parks which are concerned with various aspects of the quality of the environment. This work is likely to interest geographers, but usually requires scientific qualifications (e.g. biology, chemistry) also.

Meteorology

Geographers often study meteorology as part of their course, but ability in maths and physics is also required. Various degree subjects are acceptable, but there are degree courses in meteorology for which A level mathematics, physics and geography would be an excellent combination. There are also opportunities for scientific assistants with GCSEs at grade C or equivalent.

Oceanography

Like meteorology, this is a subject area which geographers often study. However, most oceanographers are trained scientists of various disciplines – e.g. marine biologists and chemists – and your geography would need to be supplemented with science A levels or equivalent.

Teaching

Less than one in ten geography graduates go into teaching. You must have GCSE at grade C in English and mathematics to train as a teacher. Qualified teacher status is achieved either through a BEd (Bachelor of Education degree) with geography as main subject, or a geography degree plus postgraduate teacher training. Experienced teachers might become head of department, work in teacher training, work as teachers or managers of field study centres, or do advisory/inspection work.

Library/information work

Geographers have a thorough training in researching and

handling data, which is useful preparation for work as a librarian or information officer.

Transport and distribution

Transport authorities and concerns are often very large and need logistic experts to carefully plan their current activities and future developments. The background and training of geography graduates makes them very suited to this sort of work.

Marketing and logistics

Many firms use the skills of geographers in analysing data to decide where to locate shops and offices, where and how to target their marketing, etc.

Travel and tourism

Geographical knowledge is useful in this area of work, although not essential. There are openings for people with a wide range of qualifications, from GCSEs at grade C to degrees. Employers include tourist boards/authorities, travel agents and tour operators.

Work abroad

Opportunities are varied, and can be either permanent, temporary or on a short-term contract basis. In the main, staff recruited to work on overseas projects are likely to have specialist skills and experience. The Overseas Development Administration recruits people who have expertise in land use and natural resources, for example. There are also opportunities for teaching abroad.

CARING FOR THE ENVIRONMENT

Many people are interested in the environment and concerned about the way the modern world is affecting it. A lot of occupations are concerned directly with the management of the earth's resources, and there are also jobs where you could specialise in particular environmental matters. Opportunities range from those requiring few or no formal qualifications to degree or postgraduate level.

In a sense, people with any level of qualifications, working in almost any occupation, can do their work with a concern for the environment. Caring for the environment is a way of life, whatever your job. But people with few qualifications and low levels of responsibility in their jobs are not in a position to take big decisions about the way land or other natural resources are used.

If you expect to get some GCSEs . . .

There are some technician-level, clerical/administrative and practical jobs, likely to require GCSEs at grade C or possibly an Intermediate GNVQ, in which you can play a part in the care or study of the environment – e.g. in forestry, farming, meteorology or the Civil Service. There may be opportunities for school-leavers to obtain training with an employer through the training credits system. There is a range of conservation and countryside management courses at colleges of agriculture. Most are BTEC National Diploma courses for which four GCSEs at grade C or equivalent are usually required.

The National Trust offers a training provision called *Careership* for young people wishing to work as gardeners or countryside wardens, leading to an NVQ at level 3. Students are carefully trained under full-time supervision at a specially selected National Trust centre for three years, with a total of eleven weeks of residential study each year. Applicants need to have four GCSEs at grade C in mathematics, a science, English and another subject.

The scheme is fully funded by the National Trust and sponsors, so there is no problem about transferring funding from one area to another. Only eight applicants are accepted annually to train as gardeners, plus a further eight as wardens. Six places in each category are reserved for 16–19-year-olds, while the other two are open to people up to 35 years of age. The National Trust does not guarantee a job at the end of training, but trainees are given every assistance to find a suitable vacancy.

If you expect to get A levels or Advanced GNVQ . . .

Most jobs at decision-making level are done by people with degrees or the equivalent, for which A levels, Advanced GNVQ or a BTEC National qualification are the normal entry require-ments. Suitable A level subjects for careers concerned with environmental matters are those on the science side rather than the arts: in particular, two or preferably three from physics, chemistry, biology, mathematics or mathematics with statistics, geography, geology and economics.

For some types of higher education course, mixed A levels – i.e. one or two of the above, together with arts subjects such as English, history or sociology – could be quite acceptable. Environmental science, available at some educational establish-ments as an A level subject, may not always prove an acceptable qualification for higher education courses in specialist science subjects. Look at *University and College Entrance* and the *Compendium of Higher Education* (available in careers libraries), for basic information on courses and their entry requirements.

Most professional occupations require some form of higher education, and this is where choice can prove difficult, as there are so many possible routes into environmental careers. Which you opt for will depend to some extent on your A level subjects, the degree of commitment you want to make to environmental sciences, and any specialist interests you may have at this stage.

You could consider . . .

Degree courses in fairly general subjects such as physical and biological sciences, geography, geology or agricultural economics. These could all provide a solid basis from which to branch out into an environmental career, through direct entry to a job or through postgraduate training, whilst keeping other career options open too. You will find that some institutions offer courses with a greater environmental emphasis than others, or with special options to be taken in the later stages of the course.

More specialised academic courses such as oceanography, meteorology, earth and soil sciences, deal with particular aspects of the environment and would appeal to someone who has already developed particular interests. These courses can limit general career openings and tend to merge into the next group.

Specifically vocational courses (both degrees and diplomas) providing professional training in disciplines such as town and country planning, landscape architecture, forestry, amenity horticulture, environmental health, civil engineering, agriculture and estate management. These courses require a specific commitment to the particular occupation concerned. While, as for all graduates, other general careers remain technically open, it is not always easy for someone with such a vocational degree to be accepted into, for instance, banking, primary school teaching or accountancy. On the other hand, a mandatory award is normally payable for these subjects as first degrees/diplomas, whereas finance to take vocational training as a postgraduate can be hard to come by.

Courses actually entitled Environmental Sciences or something similar. These usually involve first-year study of physical and biological sciences and perhaps some economics, sociology and human geography. In the later stages of the course, a variety of specifically environmental options are available (differing considerably from one institution to another – see the CRAC *Degree Course Guide: Geological & Environmental Sciences*). Graduates from these courses might go either directly into work or into postgraduate training (e.g. landscape architecture, surveying, planning).

EMPLOYMENT PROSPECTS

There are usually more people interested in jobs concerned with the environment, perhaps especially those involving outdoor work, than there are vacancies available. A large proportion of employment is within the public sector – central and local government, and statutory bodies such as the Environment Agency, English Nature and the Countryside Commission, which are at least partially dependent on public funding. Recent spending cuts have not increased job opportunities.

Some professionals – such as foresters, estate managers and landscape architects – will find vacancies outside the public sector, for example with practices engaged in consultancy work, or with private landowners and organisations.

Some people qualified in environmental science and related

subjects may find it very difficult to get relevant work, and it is wise, if aiming for degree-level qualifications, to choose your subject carefully so as not to close other avenues.

Openings for environmentalists also occur in the voluntary sector – e.g. the National Trust, Council for the Protection of Rural England, Friends of the Earth, Worldwide Fund for Nature. The British Trust for Conservation Volunteers and the National Trust run both weekend and week-long projects where volunteers can gain useful experience of conservation work.

Once in employment, there are National Vocational Qualifications in Environmental Conservation to NVQ levels 2, 3 and 4, which are accredited by COSQUEC (see Further Information section).

Where to look for vacancies

Advertisements for jobs with statutory bodies, private organisations and in the voluntary sector appear in Wednesday's *Guardian*, the *Independent*, and in the *New Scientist*. Some of the organisations listed in the Further Information section of this book produce newsletters with vacancies, for example the *Environment Post*.

METEOROLOGY

> Meteorology is the study of our atmosphere, from the ground to the highest levels, where it becomes 'space'. The knowledge acquired is used in aviation, planning in agriculture and industry, and in weather forecasting. There are opportunities for support staff with fewer qualifications, but meteorologists generally have good degrees.

What do meteorologists do?

Everybody knows the weather forecasters who appear on television, but there are many other opportunities in meteorology. These can be divided into four main branches:

Operational meteorology

This is the day-to-day collection and application of knowledge related to the weather. It includes the front-line personnel: TV, radio and press weather forecasters, staff who deal with telephone enquiries and staff who brief pilots and shipping. There are also many people collecting weather data from all round the world, feeding it into computers and interpreting the output, producing and analysing charts. This information is used to predict weather changes, and to provide a record of past weather, from which calculations of large-scale changes in the global climate can be made. Jobs are at various levels, with junior staff doing the routine collection and coding of weather information, and senior staff (usually graduates) interpreting and analysing it.

Research

This can involve a very wide range of projects: studying the effects of the atmosphere on aircraft; trying to find ways of intervening with the natural climate; researching mathematical techniques to improve forecasting; examining topics such as the greenhouse effect and ozone depletion.

Applied meteorology

Some meteorologists use the knowledge accumulated through observation and research in a variety of practical ways. There are uses in agriculture, forestry, fishing, civil aviation and many other industries, where meteorologists may work in a team alongside other scientists and engineers. For instance, the timing of crop spraying, prevention of flood damage, design of oil rigs and siting of power stations all involve decisions in which meteorology can play an important role. Some things really do depend upon 'which way the wind blows'!

Education

Some meteorologists work in education, in universities and other institutions where meteorology and climatology are taught. There are also opportunities to teach people involved in outdoor pursuits an understanding of the weather and forecasting. Meteorology may also be taught in schools as part of the geography syllabus.

Where do meteorologists work?

The Meteorological Office employs about half of all people working in meteorology, and is involved in operational and applied meteorology as well as research. It is an Executive Agency of the Ministry of Defence, with its headquarters at Bracknell in Berkshire, and there are over a hundred other establishments in the UK and overseas, including weather ships and stations in very remote places. Many staff work shifts, because of the need to collect and process weather information

24 hours a day. Applicants need to meet the usual Civil Service nationality requirements – basically that you should be a British, Commonwealth or Irish national.

There are, of course, also non-scientific jobs in areas like administration and computing within the organisation.

Other government departments and research councils, and the Defence Research Agencies, employ a few meteorologists, especially in research – for instance, into water pollution, building construction and radar. Various units of the Agricultural and Natural Environment Research Councils employ meteorologists, especially in those institutes which are concerned with pollution, hydrography and oceanography. Occasional openings arise with the British Antarctic Survey and the Centre for Overseas Pest Research, and with aid programmes funded through the UN. Meteorologists in these organisations work closely with scientists from other disciplines.

A few meteorologists work for industrial firms such as instrument manufacturers and geophysical contractors. There are some opportunities in the fuel industries and with off-shore oil companies. A number of firms undertake forecasting and consultative services.

EDUCATION AND TRAINING

For junior-level scientific posts, the basic requirements are usually four GCSEs at grade C, or equivalent, including English, mathematics and physical science subjects. Geography may also prove useful. Equivalent qualifications such as Intermediate GNVQ may be acceptable. These openings are advertised in the national press or in Jobcentres.

For those intending to qualify as professional meteorologists, the best way is to take A levels (mathematics and physics are especially important subjects, with geography being useful) or possibly a relevant Advanced GNVQ. You then take a degree course, which will last three to four years. There is no need to specialise too early: the degree could be in meteorology itself, or in physics (with or without meteorology options) or mathematical physics/mathematics and physics as joint honours. Some environmental science/studies degrees provide appropriate qualifications, and electronics or computing may also be useful subjects. There are also openings for holders of Higher National Diplomas or Certificates. Postgraduates are recruited into higher-level posts. The Royal Meteorological Society (see Further Information section) will provide a list of courses with a substantial meteorological content, including one-year postgraduate courses. The Society publishes a useful pamphlet, *Careers in Meteorology*.

Adults: note that maturity and previous experience may mean that stated entry requirements can be relaxed.

just.
**THE
JOB**

CARTOGRAPHY

Cartography is about making, studying and revising maps, plans, charts, three-dimensional models and globes of the earth or any heavenly body. It uses a range of sources, including the latest computer technology. Cartographers have degrees, and drafting technicians, who produce the final map, generally have A levels or the equivalent.

New technology

Cartography is a very good example of an area of work which has changed dramatically in the recent past, because of the introduction of new technology. Map-making was once a handcraft. Now, computer-assisted methods are increasingly used at every stage. For instance, sophisticated electronic equipment is used by surveyors. Computers are used to collect and manipulate data, and display the information. Most recently, the development of Geographical Information Systems (GIS) has further revolutionised map-making through acquiring information from satellites orbiting the earth. So modern cartographers have to be adaptable, and able to get the best out of ever-changing technology – but this same technology means a decrease in the number of jobs.

Where cartographers work

Cartography is quite a small area of employment, with jobs for less than 3800 people in the UK. Some firms specialise in computer-assisted or digital cartography, and these offer posts for computer personnel as well as digital cartographers.

The main areas of employment are:

- Ordnance Survey;
- government bodies and departments, such as Military Survey and the Hydrographic Office of the Defence Support Agency, the Forestry Commission, and the Meteorological Office;
- local authorities, who have technical-level jobs in their architects' and planning departments, and sections with similar responsibilities;
- commercial cartographic firms and cartographic publishers;
- service organisations like the AA and RAC;
- various industrial concerns such as oil companies;
- university geography and geology departments, which offer technician posts, as well as some teaching and research opportunities for professional cartographers.

The rest of this section takes you through the main activities involved in cartography, and explains the different jobs and training which are available.

Surveying and photogrammetry

Surveying and photogrammetry are basically quite separate careers from cartography, though of course surveys (of various kinds) provide the data from which map-makers compile their maps. (If the *surveying* aspects of the whole process interest you most, see the other sections in this book which deal in more detail with those careers.) However, surveying gets a mention here, because that is where it all starts.

Obviously, the air is the best place from which to get an overall view of a location. But you can't make a map from ordinary air photos, because photographs are 'perspective' views, in which things further from the camera appear smaller than near objects. Air photographs can, however, be turned into accurate maps if two shots of the same area, taken from different angles, are superimposed and studied stereoscopically. Radar and satellite pictures are also used to provide basic data.

Air survey work depends upon ground points having been established by land surveyors and technicians, who take very careful measurements at key points on the ground. Indeed, air survey is often regarded as a branch of land surveying. Airborne jobs are rare. Very small numbers of pilots, already holding commercial licences, are employed. There are also equally small numbers of navigators/camera operators. These are recruited from ground photographic staff and trained by their companies in survey navigation. Much of the work is overseas.

Photogrammetry describes what is done with pairs of aerial photographs, in order to turn them into maps. It involves setting up the photographs in pairs and examining them stereoscopically using binocular eye-pieces, which are adjusted until one three-dimensional image is seen. Operators record height data and topographical detail and plot graphs. Computers are used to analyse information. **Photogrammetry operators** are generally school-leavers with GCSEs or A levels/Advanced GNVQ. They are trained both on-the-job and through part-time study for BTEC National Certificates. Good eyesight is a must!

Editorial work

The professional cartographers work in the editorial office of a cartographic firm or section. Job titles include **cartographic editor**, **cartographer**, or **mapping and charting officer**, depending on the organisation. These jobs involve researching and evaluating all the information which goes towards making maps, and they normally call for graduates in cartography, geography or topographic science, and also people with postgraduate qualifications.

Cartographers do largely office-bound work. They work on the raw information which is to be used to make a map. The information they work on may be cartographic, geographic, photographic, digital or statistical. The cartographers check out its authenticity, accuracy and age, before passing it to the drafting staff who act as the production stage of map-making. The work calls for a strong interest in maps, great attention to detail, and a lot of patience. The preparation of a single map-sheet may involve several months' work.

Cartographic drafting

The drafting staff are the technicians who produce the final maps. They prepare the artwork, from which printing is done. The most important skill for drafting staff used to be ability in freehand drawing. This is now no longer the case in most jobs, as today's draughtsmen and women make great use of computer-assisted techniques. However, in some jobs, producing maps, plans or artwork for printing may still involve traditional pen and ink work, or engraving techniques.

In compiling final versions of maps, the drafting staff or technicians must comply with agreed specifications. Many maps are destined to be printed, once prepared. Various processes may be used to produce the artwork from which printing plates are made. Special artwork (whether hand-done or computer-generated) is used for maps which are printed in colour, as very many are. These need many overlays for the colour-printing process, as each main colour is put on separately.

GETTING STARTED

This is very detailed work, requiring a great deal of concentration, and good eyesight and colour vision. Drafting staff need to be technically minded and adaptable, willing and able to cope with new technology. They should preferably be educated to A level standard in geography and/or mathematics, especially if hoping to start straight from school or college. There are now some full-time diploma courses at technician level which supply most of the needs for **drafting staff**, so would-be direct entrants will find there is great competition for the jobs which do arise.

COURSES IN CARTOGRAPHY

BTEC National Certificate and Diploma in surveying and cartography

These are two-year courses, part-time for the Certificate and full-time for the Diploma. Both normally require you to have four GCSEs at grade C or above (or the equivalent), the subjects to include maths and one demonstrating fluency in English. For the part-time course, you should ideally already be in an appropriate job. There are not many full-time courses, and not all take students each year.

BTEC Higher National Diplomas

There are HNDs with titles like *geographical information systems, land surveying and mapping* or *geographical techniques*. Requirements for these courses are one A level (preferably geography), an Advanced GNVQ or BTEC National qualification, with supporting GCSEs.

Degree courses

There are degree courses with a cartographic content at a number of universities, including modular degree programmes, with titles such as *mapping sciences*, *topography* or *geographic information systems*.

Other degree courses in geography may have options in cartography or topographic science – see the CRAC *Degree Course Guide* and *University and College Entrance* for details. Most courses are normally full-time, but one or two may also be available on a part-time basis. Some are sandwich courses, with a year spent in a working environment.

Postgraduate courses

There are a few postgraduate diploma courses in aspects of cartography and topographic science. These would require you to have an appropriate first degree, such as geography.

TRAINING OPPORTUNITIES WITH GOVERNMENT DEPARTMENTS

The following government departments may offer training in cartography at technician level (between two and five GCSEs at grade C are usually specified; A levels or equivalent may be advantageous). Recruitment is very limited. Some will consider applicants in their early to mid-twenties, as well as school-leavers.

Hydrographic Office, Defence Support Agency – Admiralty Way, Taunton, Somerset TA1 2DN. Tel: 01823 337900.

Military Survey, Defence Support Agency – Elmwood Avenue, Feltham, Middlesex TW13 7AF. Tel: 0181 890 3622.

Ordnance Survey – Romsey Road, Maybush, Southampton SO9 4DH. Tel: 01703 792000. (Not recruiting staff in the fore-seeable future: recently appointed twenty new surveyors – the first for eight years. Only qualified, experienced surveyors were sought.)

School of Military Survey – contact your local army careers office for details.

LAND SURVEYING

Land and hydrographic surveyors collect the information necessary to draw up all kinds of detailed charts, maps and plans. Land surveys may be used to plan new roads or buildings; hydrographic surveys may help in the planning of new harbours or sea defences. Qualified surveyors have degrees and at least two years' practical experience. There are opportunities at technician level for those with fewer qualifications.

Land surveyor

Land surveyors are concerned with mapping. This might include remote and uncharted areas abroad, the layout of construction works, the positioning of oil rigs or the movement of the earth's surface. Land surveyors plan and organise survey work and give professional advice. They work in the field, using optical instruments, electronic meters, etc, to gauge distances and angles. This data is fed into a computer to get the coordinates they want for mapping purposes.

The few opportunities which exist in the UK are mainly with the Ordnance Survey, specialist firms, local authorities, oil/mineral companies, civil engineering firms, and public utility companies. Many qualified land surveyors work abroad. Conditions can be tough, but salaries are usually generous, with good periods of leave between projects.

Aspects of land surveying include:

Topographical surveying – plotting the features of a piece of land, using measuring instruments (theodolites) and calculations to work out the exact positioning of physical features and contour lines.

Photogrammetry – charting physical features from overlapping aerial photographs viewed stereoscopically. Most maps are compiled in this way.

Geodetic surveying – using figures obtained from widely separated locations to determine the exact shape of the earth, its current magnetic field and continental drift. Geodetic surveys are almost all done by government survey departments.

Engineering surveying – measuring features for drawing up large-scale plans showing all the topographical details of an area, and gathering information needed for civil engineering projects, such as the construction of large buildings, roads, railways, pipelines, etc.

Cadastral surveying – drawing up detailed plans of property and/or property boundaries for legal purposes, usually in connection with ownership, tenure or land registration.

TRAINING

In order to qualify as a professional surveyor, you must successfully complete an accredited degree or diploma. This is followed by a two-year period of practical training whilst working, and a final examination, by the professional body of which you are seeking membership, to test your competence. To start a degree or diploma course, you need five GCSEs at grade C, including maths and English, and two A levels. An Advanced GNVQ/BTEC National Diploma or Certificate may also be acceptable. Check with the appropriate universities for the entrance qualification required. A relevant BTEC HND/HNC may gain advanced entry to the second year of a degree or diploma course.

Graduates of other disciplines, such as arts, humanities, sciences or languages, are also recruited by surveying employers and encouraged to undertake an RICS-accredited postgraduate qualification.

RURAL PRACTICE SURVEYING

R ural practice surveying is the branch of the profession which includes land agency, rural estate management and agricultural surveying. Some people start out in this career with appropriate degrees, while others begin as trainee technicians, learning on-the-job whilst working towards professional status.

Although many land agents or estate managers work in private practice, there is scope for employment with any large landowners, such as private estate and forest owners and National Parks, and bodies like British Rail, water, gas and electricity companies, etc, local government and the Civil Service, the National Trust, Nature Conservancy and large construction companies.

Rural practice surveyors, with the appropriate qualifications and experience, can achieve membership of the Royal Institution of Chartered Surveyors (RICS) and the Incorporated Society of Valuers and Auctioneers (ISVA), professional bodies which are internationally recognised – so work abroad is a strong possibility. A real interest in and knowledge of agricultural and rural matters is vital.

Within rural practice surveying there are several specialisms. However, there is a lot of overlap between them.

Land agency
This covers the work of resident land agents, employed to manage the property of individual or institutional landowners, and

the work of those specialists in a firm of rural practice surveyors who provide a service for several estates as part of an auctioneering, valuation and estate agency practice. Land agents advise on the use, management, purchase, sale, development and value of rural land, livestock and buildings. This includes land farmed by owners, land let to tenant farmers, parkland, woodland and recreation facilities.

Rural estate management

This is the overall management of estates ranging from a few hundred acres to many thousands. The work includes the interpretation of estate accounts, and requires the legal knowledge necessary for the handling of taxation, letting farmland and houses, and rent evaluations. There can be work supervising sporting and recreational operations, with overall responsibility for perhaps hundreds of employees.

Agricultural surveying

Agricultural surveyors are more closely involved with the practical operations of farms, forests and estates, as well as the legal and financial side of things. They usually work in a practice which offers advice and supervision to a number of individual landowners, or for an organisation which owns several estates. They can advise on the suitability of different crops or livestock, their management, the necessary manpower and related accountancy, valuation and auctioneering. These skills can also be applied to buildings and machinery management.

Agricultural management

Practitioners carry out accountancy and taxation services and contribute to the day-to-day organisation of an individual farm – its livestock, cultivation, records, etc.

Forest management

Managers have responsibility for overall consultancy and accountancy for the timber production of woodlands – organising the establishment of young plantations from seedlings, and their maintenance and protection from fire, pests and diseases, to the point of marketing the timber to sawmills.

Auctioneering, valuation and claims

Skill in valuation is essential in dealing with country property and produce. Surveyors may be concerned with the auctioning of farms, houses, timber, livestock and chattels. Other duties are recreational land management, dealing with rural property as an investment, legal work and planning. Individuals often specialise in one area, such as rural planning, leisure and conservation, farm business, administration, tenants' rights, valuation, compensation, etc.

Technical surveying

It is possible to work as an agricultural technical surveyor, acquiring wide experience in valuation, land drainage, reclamation, farm buildings and management. Training is through a combination of practical experience and coursework, studying at college for a BTEC National, and then Higher National Certificate or Diploma in Land Administration (Estate Management).

TRAINING

See the previous section about training to qualify as a professional surveyor.

To study for the examinations of ISVA – the professional society for valuers and auctioneers – you need a minimum of five GCSEs at grade C, including maths and English.

Lists of full-time, part-time and correspondence courses, and accredited degrees and diplomas, are available from the RICS and ISVA (see Further Information section).

BUILDING SURVEYING

Building surveyors are the kind of surveyor you are most likely to have met, particularly if you have moved house recently. They are experts in building construction, maintenance and repair and have usually taken a degree in building surveying. Some reach professional status through combining job experience with college training.

Building surveyor

One of the building surveyor's jobs is surveying houses for clients who are thinking of buying properties, and for building societies or banks who have been asked to lend money for the purchase of a house. The surveyor examines the structural soundness of the building. Is it well built? Has it been properly maintained? Are the drains, electrical wiring and plumbing adequate? Is the roof sound and weatherproof? Are there any signs of dry rot or woodworm? To do this competently requires considerable training and experience, particularly as, under the law, a surveyor can be held liable for mistakes. A discontented client can sue a surveyor for expenses incurred due to faults the surveyor should have noticed.

Building surveyors also:

- advise on the construction of new buildings and the alteration and repair of existing ones;
- handle all the legal aspects of planning, site surveys and layout of new developments – private, industrial and commercial – including overall control of a project from beginning to end;

■ may be employed by organisations such as the National Trust to advise and manage the restoration and refurbishment of historical buildings.

The job combines both practical and administrative work. Depending on the type of firm they work for, surveyors could find themselves inspecting drains and lofts one day, and writing reports or addressing a planning meeting the next. Most jobs involve some outdoor work and all involve desk work. It is important to be able to present clear, concise reports in writing, and to present information well verbally. Getting on well with a wide range of people is also vital.

Technical surveyor

Technicians are also employed on all aspects of the work, to carry out functions which do not require the knowledge of a fully-qualified surveyor.

Where do building surveyors work?

Employment can be with private practice surveyors and specialist firms of building surveyors – a young surveyor could hope eventually to become a partner in a practice. Some firms in private practice are large, employing many surveyors; others are small and may offer opportunities for a qualified surveyor to buy a partnership.

A wide variety of other organisations also employ building surveyors. They include local and central government, nationalised industries, property companies, banks – in fact any organisations that own or administer large numbers of buildings and thus need surveyors.

TRAINING

In order to qualify as a **chartered surveyor**, you must successfully complete an RICS-accredited degree or diploma. See details in the section on land surveying.

Technicians can train through full-time or part-time BTEC courses, requiring four GCSEs at grade C, or equivalent, to include English and maths. Those with lower qualifications may still be eligible to train over a longer period. The Society of Surveying Technicians can give further information (see Further Information section).

EMPLOYER–BASED TRAINING

There is a London-based scheme through which trainees can study for BTEC qualifications. School-leavers from other areas can apply, but need to have registered for funding at their local careers centre.

Civil engineering surveyor

Civil engineering surveyors support the work of civil engineers throughout the construction industry. Their skills are required both for multi-billion pound projects, such as the Channel Tunnel, and for small-scale operations, such as the renewal of a few hundred metres of service pipeline. **Quantity surveyors** (see next section) provide the financial and contractual control; **land surveyors** draw up the plans for a project, showing topo-graphical details and any additional information needed for the construction work, monitoring the progress on a day-to-day basis. There are degree courses in civil engineering surveying or, alternatively, you can qualify through BTEC National and Higher National Diploma courses and through the Institution of Civil Engineering Surveyors' own examination structure.

For a degree course, you need a minimum of two A level passes, or a good grade in a qualification of equivalent standard, such as the Advanced GNVQ in Construction and the Built Environment.

QUANTITY SURVEYING

Q uantity surveyors are experts in the cost and management of construction projects. They may work in private practice, for a building contractor or for commercial and industrial companies. Most quantity surveyors have degrees, although there are also opportunities at technician level.

Quantity surveyors have been described as the economists of the construction and civil engineering industries. They act as a link between the architect and engineers on one side, and the contractor on the other. They are key people, involved at all stages of a project. A major part of the cost of any building or civil engineering project is represented by the materials to be used. How much steel or concrete is needed to build a motorway flyover? What price can a particular type of steel be obtained for? Who can supply it? In calculating the cost of a project, many factors of this sort must be taken into account.

What the work involves

A quantity surveyor . . .

■ works with a contractor to produce initial plans and drawings for a project, and helps to calculate the most economical way to do the job;

■ assesses the 'quantity' of labour, materials and equipment required to complete the project;

■ prepares a 'Bill of Quantities' from which an estimation of

the entire cost of the project can be drawn up by the contractor;

■ assesses interim payments to be made for materials;
■ follows the progress of the project to make sure it is completed on time and within the budget.

What it takes

A quantity surveyor needs:

- a good head for figures;
- a sound knowledge of construction;
- knowledge of industrial and contractual law;
- organisational skills;
- communication skills.

Where do quantity surveyors work?

Quantity surveyors are employed in large contracting and civil engineering firms, in local and central government departments and in private practice, such as specialist firms of quantity surveying consultants. More than half of all quantity surveyors work in private practice. People who like the outdoor aspect of the work will find a job with a contractor the best alternative.

Steve – quantity surveyor

'My job is largely about getting good value for money. I work from the architect's plans, calculating how many bricks, tiles, sacks of cement and so on are going to be needed for a job. I need to know about the uses of different materials, where to buy the best quality at the best price, and so on. I also have to predict how many craftspeople will be needed for how long, and then work out how much the whole job is likely to cost from start to finish. I work with the architects, the builders' merchants and the builders themselves, so I get to know a lot of people in the industry. It's a very responsible job, because if I get my sums wrong, it can mean the difference between profit and loss!

The work is a mixture of desk duties and site visits, which means that every day is different. I enjoy the hard hat *and wellie days, but sometimes it's nice to be able to retreat to a cosy office.*

To qualify, I first had to get good GCSEs and then two A levels (an Advanced GNVQ in Construction and the Built Environment might be an alternative to go for now – it wasn't around when I was at school) before going to university to study for three years. I became a student member of the Royal Institute of Chartered Surveyors while on my degree course, and after graduating and working for a year was able to qualify as a chartered surveyor.

I'm now qualified to work anywhere in the European Union. Maybe, one of these days, I'll get a job where they're building a luxury holiday villa complex somewhere on the Mediterranean!

TRAINING

In order to qualify as a **chartered surveyor**, you must successfully complete a degree or diploma which has been accredited by the Royal Institution of Chartered Surveyors (RICS). See details in the section on land surveying.

To become a **technical surveyor**, you need four GCSEs at grade C, including mathematics and English. You can train either whilst you are in employment, or full-time at an FE college. Further details can be obtained from the Society of Surveying Technicians (see Further Information section).

Professional examinations in quantity surveying are also offered by the Association of Building Engineers, the Architects and Surveyors Institute, and the Chartered Institute of Building.

MINING SURVEYING

Mining or minerals surveying is a unique and specialised part of the surveying profession. The job combines practical surveying work, both above and below ground, with periods at the drawing-board. Nowadays, surveyors have usually started with a degree or HND and achieved professional status with job experience. There are some opportunities to start at trainee technician level.

Every mine in the UK has to have a *'surveyor of the mine'* – someone who is legally responsible for the accuracy and completeness of the plans and drawings of those particular mine workings. This is essential in the interests of safety. The surveyor must prepare comprehensive plans, drawings and sectional diagrams of the mine, surface workings and geological formations. The plans must show boundaries, outlines and levels of neighbouring mines and abandoned workings. Plans must also be kept of electrical and ventilation systems. The law requires that records and surveys have to be updated every three months.

The actual surveying work is similar to that done by a land surveyor, making use of sophisticated electronic surveying and measuring equipment. *Remember that underground conditions can be tough.*

As well as this basic job, **mining** or **minerals surveyors** can also be involved in the acquisition of working rights, and assessing such things as loss of land value because of mining activity and the extent of damage to an area through subsidence. They

advise on reclaiming and infilling land after minerals have been extracted.

EMPLOYERS

Quite a number of minerals surveyors in this country are

employed by private coalmining companies, such as RJB Mining. Others work for firms which mine or quarry other minerals, such as salt, clay, gypsum, tin, fluorspar and iron. The minerals section of the Inland Revenue employs a few surveyors, and there are occasional posts with local authorities and organisations such as British Rail and the Forestry Commission.

There are good opportunities overseas for qualified surveyors – in gold mines or in extraction operations involving minerals such as bauxite, uranium, precious stones, etc.

EDUCATION AND TRAINING

To train as a **minerals surveying technician**, four GCSEs at grade C (including maths, English and science) are required.

There are degree and HND courses in mineral surveying offered at various higher education establishments. In order to qualify as a **professional surveyor**, you must successfully complete an accredited degree or diploma. This is followed by a period of practical training whilst working, and a final examination, by the professional body of which you are seeking membership, to test your competence. There are also accredited postgraduate qualifications for arts, sciences, languages and humanities graduates. Check prospectuses for exact entrance requirements, and with professional bodies for up-to-date information on accreditation.

MINING & MINING ENGINEERING

Jobs in the mining industry range from hard physical work using modern automated machinery, through engineering and surveying work to commercial sales and marketing. The industry employs people with qualifications ranging from a few GCSEs to postgraduate level. All entry routes are now open to men and women.

Deep-mine coalmining has declined sharply in the UK. Opencast mining is also falling off, but at a slower rate. Together, they now represent a much reduced part of the minerals industry. There is a small amount of extraction of minerals such as tin, salt, fluorspar, potash and gypsum, and larger extractions of aggregates, limestone, china clay, etc.

There are very small pockets of metallic ores in Britain, not worth exploiting on the commercial scale. However, mining of metal-bearing rocks takes place on other continents, such as Africa, North and South America and Australia, where zinc, copper, silver and platinum are extracted and purified. British-trained mining personnel are to be found in every continent and many new mining graduates find their first jobs overseas. The Mining Association of the UK (see Further Information section) represents worldwide operations and can provide a list of its member companies.

Metalliferous and miscellaneous mining and minerals processing

This is a small industry in Britain. Few people are recruited, and there are no well-defined entry schemes for young people. However, graduates of mining disciplines gain a lot of technical proficiency, applied computing skills and some business management training during their coursework, which opens up opportunities in other fields – for example, quarrying, tunnelling, commodity broking, mining journalism and environmental management.

Most mining and mineral processing engineers are recruited as graduates from the BSc/BEng courses in mining engineering, minerals processing and engineering, and minerals processing, available at the universities of Leeds and Nottingham, Imperial College, London, and the Camborne School of Mines at Redruth (part of the University of Exeter), which also offers a two-year, full-time BTEC HND course in either mining and

minerals engineering or minerals surveying and resource management. The metalliferous mining industry has, in the past, supported a scholarship and bursary scheme for graduates studying mining engineering or minerals processing/engineering. Details of the current position are available from the Mineral Industries Educational Trust.

QUALIFICATIONS

For entry to a degree course, a minimum of two A levels in mathematics and physics are required, plus supporting GCSEs. For entry to BTEC Higher National Diploma courses, four GCSEs at grade C and one science A level are required, together with evidence of completed course work for another A level subject. A BTEC National qualification in a relevant science or engineering subject, or an Advanced GNVQ, may be acceptable alternative entry qualifications. Check with individual institutions.

Coalmining

The operations of British Coal have been effectively taken over by private companies, and recruitment and training have been very limited during the changeover. The largest of these companies is RJB Mining, which is now recruiting craft apprentices in electrical and mechanical engineering and graduates in mining subjects. Celtic Energy Ltd is now operating the majority of coalmines in Wales, only one of which is a deep mine. Some training placements are offered to young people, but no further recruitment is currently taking place.

What the work involves

Coalmining is a much safer job than it used to be – health hazards have been greatly reduced, and most of the really back-breaking work is now done by machinery. But it is still a harsh working environment, and serious accidents can and do happen.

Conditions can be claustrophobic and hot, and there are extensive stretches where it is necessary to crawl along underground, and where walking is rough. These conditions affect management and engineering staff as well as the miners.

Opencast mining provides a working environment which is more like a quarry (see next section). All opencast mining is performed by private contractors.

Levels of work
Craft and operative work
Mining and engineering craftworkers and operatives work in both surface and underground operations, under the supervision of professional mechanical, mining and electrical engineers. Miners are generally local people from the pit areas.

Technicians, and incorporated and chartered engineers
Technicians are normally people with BTEC National qualifications or equivalent. Higher National qualifications lead to **incorporated engineer** status. **Chartered engineers** are almost always graduates, and they are the managers of the coalmining industry, responsible for the overall running and exploitation of a mine.

There is a need for great technical expertise and knowledge, as well as the ability to manage the workforce. To produce coal efficiently, safely and profitably, there must be a thorough understanding of basic engineering principles as well as of relevant aspects of surveying, geology, soil mechanics and explosives. Mining engineers, electrical, electronic and mechanical engineers will continue to be employed in the new privatised companies, and these may well offer sponsorships and scholarships in the future.

Graduates of computer science, fuel science, chemical engineering and other related subjects are also employed.

QUARRYING

Quarrying provides us with aggregates — crushed rock, sand and gravel — to construct, improve and maintain our homes, workplaces, hospitals, schools, and most other structures which make up our built environment. As well as aggregates, granite, limestone and other rocks are quarried. There are opportunities for engineers and geologists as well as for those with practical and mechanical skills.

Quarrying is an ancient industry dating from people's realisation that they could make themselves homes from stones, rather than having to rely on finding unoccupied caves. Stone was laboriously hacked, split and heaved from the rock face and sometimes taken many miles to its final destination.

Nowadays the industry is still very important. Although we rarely build our houses entirely from stone, we still need crushed stone for roads and concrete-making, clay for bricks and pottery, limestone and gravel, slate and sand. All these materials are needed in large quantities by the construction industry and for various manufacturing processes. However, we no longer rely on strong arms to extract these materials. Most quarrying is now achieved through the controlled use of explosives, followed up by enormously powerful machines.

This means that, now, only about 30,000 people are involved in British quarrying. That figure includes not just quarry workers, drillers and shotfirers, but managers, geologists, surveyors and office staff. Whether you live amongst the hard rocks of the north and west or the clays, chalks and gravels of the south and east, there is very likely to be some quarrying going on in your area. Quarries are located throughout the country, often in rather isolated places. Although there are some very large quarrying firms, each individual quarry usually has only a small number of employees. Few companies employ more than ten people; only 120 employ fifty or more people.

Working in a modern quarry

Modern quarrying is a largely mechanised business, though it is still necessary to be fit and active. The workers sit at the controls of large and powerful machines which process huge amounts of material every day. Quarries are noisy, dusty places, and dangerous if safety rules are not kept. Much of the work is outdoors and, although modern machines are normally equipped with weatherproof (and soundproof) cabs, you may have to get out of them quite frequently. Machines must be repaired and adjusted, regardless of whether it is dry and warm or cold and wet.

Machine operating

Trainee operators are sometimes taken on by quarrying firms. There are some restrictions because of Health and Safety legislation and local by-laws. Operators can become skilled at working the more complicated machines, although the work remains basically routine. Training is mainly on-the-job, but operators may be required to go on short courses in plant and vehicle operation.

Engineering and maintenance

Expensive machines cannot be left idle when they break down, so there are jobs in quarrying on the maintenance side for various skilled workers. Such jobs include plant mechanics and fitters, vehicle mechanics, electricians, welders and tuners. These

positions may be filled by workers who are already trained, or by young people with GCSEs in English, maths and science, who are occasionally offered apprenticeships lasting about four years.

Technician, supervisory and management posts
Technician
Technicians can work either in laboratories or in the field, in quality control, surveying or engineering. School-leavers with four GCSEs at grade C, including English, maths and science, may be taken on to study part-time for a BTEC National Certificate, or there may be direct entry to the industry for people who have already followed a two-year full-time college course to gain a BTEC National Diploma.

Managers and supervisors

Known as foremen and women, these are the people responsible for the daily operation of the quarry in meeting set output targets. They make sure that everything works smoothly – that the machinery is functioning and there are no staff problems. They may also be involved in waste disposal and the infilling of abandoned workings in restoration programmes. Some supervisors and managers are promoted from machine operator or craft-level jobs when they have gained experience; others may start by taking a technician-level course. Managers may also have a responsibility for on-site processing of the extracted minerals, for example running a concrete-producing plant.

There is a three-year sandwich degree (BEng) in Quarry and Road Surface Engineering at Doncaster College, which covers the exploration and assessment of new quarry products, materials testing for quality control, the design of open pits, and hydrology – understanding the water cycle and the water tables in different geological localities. BEng Honours degrees are available at the University of Plymouth and Imperial College, London.

Civil and mechanical engineering or geology-based degrees and Higher National Diplomas are also useful. Larger companies may also take on graduates or HND holders in business studies, economics or environmental studies for management training. Some employers offer sponsorships for degree courses, but you would need to check with individual employers.

The University of Surrey offers a BSc Honours degree in Composite Materials Technology.

There are suitable short, part-time and correspondence courses for technicians and trainee managers, including those run by the Institute of Quarrying, City & Guilds and Doncaster College. Specialist courses are offered in certain aspects of the work, such as safety, blasting, production efficiency and environmental control.

ESTATE AGENCY

Buying a house is often the largest investment we make in our lifetime. We depend upon estate agents to help us through the process. Most of their work is concerned with buying and selling properties on behalf of someone else. Although it is possible to become an estate agent without any educational qualifications, senior negotiators and branch managers generally have degrees, or the equivalent, plus professional qualifications.

What it takes

To work in estate agency you need to:

- write descriptions of properties;
- learn about the legal aspects of property work;
- be able to drive, and be prepared to do a good deal of travelling;
- like meeting people;
- look smart;
- be adaptable – you will have dealings with all sorts of people;
- be a good communicator – property transactions can be complicated;
- work in the evening and at weekends – increasingly, estate agents have to work when their clients are free;
- work under pressure – your earnings may depend on how much you sell;
- be patient and tolerant – it takes a while to complete property transactions, and many deals fall through.

Working as an estate agent
Buying and selling property
Much of the work is concerned with the sale of land and houses. The seller pays for the services of the agents, who value the property to decide an asking price, arrange newspaper advertisements and prepare handouts, photographs and perhaps even videos to show to likely purchasers.

Some estate agents specialise in dealing with commercial properties, such as office blocks, factories, shops and pubs. Valuing and selling such properties are much less straightforward than dealing with private property, as factors such as customer goodwill have to be taken into account.

Selling property involves legal transactions, so estate agents regularly liaise with solicitors or licensed conveyancers, and need to have a good understanding of property law themselves.

Investment work
Clients may be private individuals, large property groups or insurance companies. Agents need to know the property market very thoroughly to advise on the best type of investment for particular clients. Negotiations can be very lengthy.

Management of property
This involves approving leases, selecting tenants and supervising maintenance work, and needs some knowledge of lifts, heating/ventilating systems, etc. Accounts, rating, tax and insurance are also dealt with.

Other jobs
These can include valuation and auctioneering, negotiations over community charge assessments, planning matters, compulsory purchase compensation, inventories for furnished lettings, rent collection, etc. In rural practices, tenant rights' valuations and arbitrations over farm rents can be important issues.

Working as a negotiator

Negotiators are employed by many estate agents to do the less complicated work. They work on house sales, rather than on the more complex industrial and commercial work. They deal with purchasers, show them around houses and help clients to find a suitable property. House purchase usually involves a chain of sales and purchases, and so the negotiator has to keep an eye on the progress of the sales, trying to keep everything moving along.

GETTING STARTED

Negotiators often have no professional qualifications. They can start in a firm as a receptionist or clerical assistant, gradually taking on negotiating work as they learn the business. Negotiators can, however, decide to gain a professional qualification from either the Royal Institution of Chartered Surveyors (RICS), the Incorporated Society of Valuers and Auctioneers (ISVA) or the National Association of Estate Agents (NAEA). The College of Estate Management offers a correspondence course in residential estate agency which requires no specific academic qualifications, only motivation and a good command of English. This might satisfy some of the academic requirements for membership of the Society of Surveying Technicians or NAEA. There are often opportunities for good negotiators to become office managers or associate partners.

It is possible to start as a **trainee negotiator** through employer-based training opportunities. Your careers centre will know if there are any local opportunities.

GETTING PROFESSIONAL QUALIFICATIONS

There are no compulsory standards for entry, but this may change through public pressure to allow only registered practitioners to practise. Education and training should lead to membership of one of the following professional bodies:

Royal Institution of Chartered Surveyors (RICS)
In order to qualify as a chartered surveyor, you must successfully complete an RICS-accredited degree or diploma. See details in the section on land surveying.

Incorporated Society of Valuers and Auctioneers (ISVA)
ISVA is the professional society for valuers and auctioneers. Minimum entry requirements are five GCSEs at grade C, including English and mathematics. There are various ways of acquiring ISVA qualifications: three-year full-time courses, approved employment plus part-time study, distance-learning

courses through the College of Estate Management while in approved employment, or exemption through a degree or other approved qualification. ISVA publishes a free list of exempting qualifications.

National Association of Estate Agents (NAEA)

There are no stringent entry standards for basic membership, which is normally open to people practising as estate agents. At Corporate grade level, however, there are compulsory entry requirements – the successful completion of a test is required for the grade of Associate, and the Fellow grade is only open to agents who have attained the NAEA Intermediate Certificate of Practice in Estate Agency (CPEA) or a comparable qualification. The CPEA is the highest qualification in estate agency, for **senior negotiators** and **branch managers**.

The NAEA offers correspondence courses in residential estate agency and estate agency practice, and short courses – e.g. one-day courses in residential lettings, valuation, building construction, etc. They also recognise the correspondence course in residential estate agency run by the College of Estate Management in Reading.

NVQs and Modern Apprenticeships

There are NVQs up to level 4 available in Residential Estate Agency. The Residential Estate Agency Training and Education Association (REATEA) can give further details.

Modern Apprenticeships in Residential Estate Agency are available to young people aged 16 to 19 who have a minimum of four GCSEs at grade C. A Modern Apprenticeship lasts approximately three years and offers training up to at least NVQ level 3. Contact your careers service or Training Enterprise Council (TEC) for details of opportunities in your area.

VALUATION & AUCTIONEERING

Valuation is about estimating what something is worth; *auctioneering* is about selling it for a good price. There is a lot of scope for specialising in particular aspects of valuation and auctioneering, and a wide range of employment opportunities. It is possible to start training with GCSEs, and work your way towards professional qualifications.

Valuation

Property of any kind has different values for different purposes. A machine, as a piece of engineering equipment, may be worth thousands of pounds. As scrap metal, or for insurance purposes, it might have quite different values. Valuers are experts at assessing these values.

Many valuers are experts in real estate – valuing buildings and land. A property might be valued for a mortgage application, or for a seller or purchaser, or for insurance purposes. To arrive at a figure, the valuer takes into account things like the size, age, appearance and condition of the property, where it is, and what prices similar properties are fetching.

Some valuers concentrate on antiques and works of art, where a good deal of experience is required. There is plenty of opportunity to specialise.

You can also get involved in valuation alongside other aspects of your work. For instance, all residential estate agents have to be able to value straightforward properties.

Auctioneering

Auctioneering is about selling. Things which are sold by auction fall into three groups: land and buildings, livestock, and chattels – which means any sort of moveable goods, ranging from furniture and household belongings to cars, machinery and works of art.

Auctioneering starts with gathering things to sell. Items from many different sources might be brought together for a general sale of furniture. Sometimes, a firm of auctioneers might hold sales of particular items like stamps, steam engines or antique dolls. Getting hold of items to auction requires skill and knowledge on the part of the auctioneers.

People doing this work have to know a lot about valuation and must be able to sense how the bidding is going and get the best possible price for a client. Contrary to popular opinion, they don't have to be able to talk at a thousand words a minute, though things do move very quickly at some types of auctions.

As with valuation, this sort of work is not done only by specialists. In a small town practice, for instance, where auctions are not held very often, they are taken by the partner or member of staff with the most suitable personality and experience.

Where valuers and auctioneers work

Private practice in estate agency, auctioneering and surveying

A private practice usually does a wide range of work – few firms deal only with valuation and auctioneering. The work depends to a large extent on the location of the firm.

In a rural practice, much of the valuation work involves property – land, farms and stock – for letting and sale. Valuers advise on rents and value farms, crops, stock, etc, for taxation purposes. They value farms when tenants leave.

Valuers in town and city practices specialise in offices, shops, residential property, factories and industrial sites. Valuations may be for insurance purposes, sales and purchase of property, settlement of disputes over property rights and assessments for income tax. A typical job is valuing a shop, bearing in mind its position and possible takings.

Education and training should lead to membership of one of the professional bodies mentioned later in this section.

Financial institutions

Valuers can also be employed by financial institutions such as banks and building societies and property companies. These posts would generally be for qualified people.

Fine art and antiques

Some firms (or sections of firms) specialise in valuing and auctioning antiques and works of art. They include the famous firms of Sotheby's and Christie's which, not surprisingly, are very competitive to get into.

Expertise in this area is built up through experience, but there are courses which can help to prepare you for this sort of work. Private courses are offered by Sotheby's and Christie's, or there is a three-year BA Honours course at Southampton Institute of Higher Education. Entry is with five GCSEs at grade C, or equivalent, to include mathematics and English, plus two A levels. There is also a two-year graduate entry course.

The Civil Service

The Valuation Office of the Inland Revenue is the country's largest employer of professional surveyors and valuers. Professional staff make valuations of property for tax purposes – when people inherit, sell or transfer property. Another major aspect of the work is valuing property for acquisition, negotiating leases, and renting and disposing of land and buildings.

Valuations are carried out on behalf of the government, local authorities and other public bodies.

The Property Services Agency of the Department of the Environment also employs valuers to deal with the valuation, acquisition and disposal of all types of property on behalf of the government.

Local government
The knowledge which members of the Institute of Revenues, Rating and Valuation have, regarding levying and recovering

revenues, is sought by departments in local authorities, as the community charge is based on property values. Valuers also often advise their authorities on valuation and management of properties, rents and purchase prices, as well as landlord/tenant problems, and rent rebate schemes. The rateable values of properties are monitored by valuers in private practice or in local government. These valuers deal with appeals and negotiate with the **valuation officers**.

Normal minimum requirements are four GCSEs at grade C, to include English and mathematics, plus one A level, for entry to the exams of the Institute of Revenues, Rating and Valuation. Graduates with appropriate degrees, together with two years' post-qualification experience, are exempt.

QUALIFICATIONS IN VALUATION AND AUCTIONEERING

Although there is no legal requirement, it is usual to work for the examinations of one of the professional bodies concerned with auctioneering and valuation. Firms with a good reputation will certainly insist on their staff becoming properly qualified. Which professional body you choose to join will depend on factors such as your academic qualifications, the type of work you do or hope to do, and the traditions of the firm you work for.

There are various ways of studying for professional exams (full-time, part-time or home study) and various starting points (after GCSE, A levels/Advanced GNVQ or a degree): see the section on full-time and sandwich courses, below. Normally, to be eligible to take part-time or correspondence courses, you have to be in approved and relevant employment.

Royal Institution of Chartered Surveyors (RICS)
In order to qualify as a chartered surveyor, you must successfully

complete an RICS-accredited degree or diploma: see the section on land surveying for details.

Incorporated Society of Valuers and Auctioneers (ISVA)
This is the professional society for valuers and auctioneers. Entry requirements are a minimum of five GCSEs at grade C, to include English language and mathematics. Some relevant degree courses can exempt you from the ISVA examinations or enable a special qualifying exam to be taken. Full-time, part-time and correspondence courses are offered for ISVA exams.

Incorporated Association of Architects and Surveyors
Entry requirements are a minimum of two A levels and three GCSEs at grade C, to include English, mathematics and science. There are exemptions for appropriate degrees.

Full-time and sandwich courses
There are relevant three- and four-year degree courses (e.g. in estate management). These need at least two A levels plus three different GCSEs at grade C for entry, to include English and mathematics. Suitable BTEC qualifications and other alternatives are also acceptable.

Various BTEC Higher National Diplomas (two- and three-year courses) are also offered. Entry requirements are one A level and four GCSEs at grade C, to include English and mathematics, or a suitable BTEC/GNVQ qualification.

The professional bodies publish lists of accredited courses.

Adults: note that maturity and previous experience may mean that stated entry requirements can be relaxed.

LANDSCAPE ARCHITECTURE, MANAGEMENT & SCIENCE

L andscape architects, managers and scientists create land-scapes to make the area we live in a pleasant place. They may design public parks, reservoirs, the surroundings of public buildings or private gardens. What they design must fit in with the environment. Landscape architects have degrees or the equivalent. Technicians need at least good GCSEs.

Landscape architect

Landscape architects can work on settings for all sorts of build-ings, roads, parks and play areas, and other construction pro-jects, such as housing developments. The aim is to do the least possible damage to the scenery and ecology of the area. Landscape specialists often work on projects such as preserving the coastline, rescuing sites which have been left derelict, and restoring disused pits and quarries. Landscape architects don't just work in the countryside – nowadays they are more and more concerned with making our towns and cities nicer places to live in.

What the work involves

Whatever the particular project, there are similar things to be done. Landscape architects hold discussions with their clients, to find out what the particular job is all about. They make visits to sites to carry out surveys, and then draw up plans and designs. These have to be agreed with the client. There is a good deal of paperwork to do. The detailed plans have to include projected costs for the project, including cover for maintenance work after the work is completed!

Once the designs and costs are agreed, and contracts signed, the work can begin. The landscape architect visits the site from time to time, to check that the work is progressing smoothly and as agreed in the contract.

The landscape architect is basically a designer, but, in order to be able to produce a design, he or she must have an understanding of related topics such as civil engineering, surveying, geology, horticulture, earth-moving techniques, etc.

Teamwork
If you have got the idea that landscape architects work alone, then you are very much mistaken! They are usually part of a team which includes architects, civil engineers, town planners and the construction site foremen. A lot of time is spent at meetings and discussions and it's important to be able to work well with people at all levels.

Indoors and out
Qualified landscape architects usually spend less than a quarter of their time outdoors. The major part of their time is spent on desk-work and meetings. Landscape architects working in private practice are likely to spend quite a lot of their time travelling, undertaking commissions around the country.

Landscape scientists and managers
Landscape scientists are concerned with the physical and biological aspects of a designed landscape. Decisions regarding the best method of reclaiming a site, or the most suitable plants to cultivate, often need their scientific expertise. To be a landscape

scientist, you need a suitable degree (perhaps in botany or ecology) followed by relevant practical experience and/or a higher degree in a subject such as soil science, geomorphology, etc.

Landscape managers are concerned with the long-term management and development of schemes and projects. Different uses of land can require careful management, where, for example, there may be a conflict between recreation and conservation needs. The manager will also see to any practical maintenance needed to keep a design in good order. Forestry and farming skills may be involved in this sort of work.

Reading University offers a first degree course in landscape management accredited by the Landscape Institute. There are other suitable first degree and diploma courses in landscape management, land management, horticulture, agriculture, forestry, geography or botany. Also, there is an accredited post-graduate course at Wye College, University of London. If you follow an appropriate, but unaccredited course, you will need a few years' relevant experience before becoming qualified.

What it takes
To be a landscape architect you need:

- a concern for the environment;
- an understanding of conservation;
- creative ability;
- the ability to draw well;
- good communication skills;
- a practical outlook;
- technical and scientific understanding;
- imagination and enthusiasm.

WORK OPPORTUNITIES

Landscape architects, managers and scientists can work in public service or private practice. Approximately half work in public service: that means central and local government, new town development corporations, water authorities, etc. The other half are mainly in private practice, or salaried posts with large industrial concerns. A few go into teaching or lecturing. Private practices vary in size from a single practitioner, working on their own, to large specialist firms. Other professionals, for example architects, may also have landscape architects working in practice with them. There are some opportunities to work overseas, mainly in the Middle and Far East.

EDUCATION AND TRAINING

Associate membership of the Landscape Institute (ALI) is the recognised professional qualification in landscape architecture. To achieve this status, you must pass the Institute's Part Four (professional practice) examination, after you have attained graduate membership and had at least two years' practical experience.

People who want to become landscape architects can start their careers in one of two ways:

- **by taking a degree in landscape architecture:** the minimum requirements are two A levels or their equivalent, plus supporting GCSEs which should include English and either mathematics or science. At A level, the Landscape Institute regards subjects like art, biology, botany and geography as particularly relevant.
- **by further study after taking a degree in a related subject** (e.g. architecture, horticulture, botany): the higher degree can be taken straight away, or after some time has been spent in related work.

All applicants should carefully check the specific requirements of courses they are interested in.

Adults: mature applicants may be able to gain exemption from normal minimum requirements if they have relevant experience, for example at technician level in architecture, horticulture, botany or forestry.

During training, the subjects studied include: graphic design and drawing, history of art, ecology, geology, horticulture and botany, architecture and building construction, planning and planning law, surveying and site assessment, techniques of earthmoving. Remember that not all of these subjects will be studied in depth. For instance, although horticulture and botany are important to the landscape architect, he or she will not study these to the same level as students on specialised single-subject degree courses.

Courses

There are full-time courses leading to first degrees and postgraduate degrees and diplomas at a number of universities and colleges. Some are based in art faculties. For up-to-date information on the courses offered, consult *University and College Entrance* and the Laser *Compendium of Higher Education.* You could use the courses information service database offered by most careers centres – ECCTIS 2000. Some part-time postgraduate and undergraduate courses are available. Applicants are advised to check whether the course they are interested in is accredited by the Landscape Institute.

Technician and craft-level work

There are some courses offered at technician level, on a full- or part-time basis. Examples of courses available include the BTEC National Diploma in landscape construction and college awards in landscape practice or design. These generally require some GCSEs at grade C for entry. Craft-level qualifications, such as City & Guilds, are available for those looking for practical courses. GNVQ in Land-based Industries could also provide a starting point.

There are jobs at both levels with landscape contractors, local authorities and large landowners. Pay and conditions are similar to those in agriculture and horticulture.

The British Association of Landscape Industries can provide a list of its members in each region.

TOWN & COUNTRY PLANNING

Planning the 'town and country' varies in scale from national government decisions about roads and power stations to local government decisions dealing with house extension applications. Planners also have to create a balance between the 'built' and the natural environment. Town and country planning is a profession for those with a degree, and there are also opportunities for planning support staff with fewer qualifications.

Town planning is a very broad area of work. Whilst mainly being concerned with land use, planners have to know about many related areas of work such as architecture, archaeology and nature conservation. It is, however, a profession in its own right and the Royal Town Planning Institute has over 17,000 members. The work is varied and may include undertaking social surveys, public meetings, investigating the effects of different traffic schemes, evaluating alternative uses for vacant sites, the promotion of schemes to attract industry to an area, and formulating land reclamation programmes. As planners progress in their career, it is possible to specialise in particular aspects of the work.

What planning is all about

Planners have to think very carefully about the environment. Planning is for people in the community who often want to have some influence over the area in which they live. Television often carries stories of local disputes over the siting of a new road or supermarket. If projects are likely to have a significant effect on the landscape, the developer has to carry out an assessment of the environmental impact.

Planning law is administered by central and local government, although many other bodies are involved. The Department of the Environment is responsible for approving overall structure plans for county areas, commenting on local plans and producing guidance on specific planning policy – for example in relation to highways or archaeology. In Scotland, planning law is administered by the Scottish Office Development Department in Edinburgh

County and district councils

Currently, only county councils prepare the 'structure plans' which set out broad policies for their area, but the new unitary authorities will assume that responsibility, probably in a joint capacity with other councils. These plans are based on detailed surveys of the county and surrounding region. The county council must publish their plans and take account of the opinions of the public and other bodies before the plan is adopted.

Local plans prepared by district and borough councils show in greater detail how the proposals of a structure plan will affect a particular area, and contain detailed schemes for different areas and their use – e.g. shopping, housing, agriculture, industry or recreational facilities. Social, economic and environmental factors have to be balanced, and the decisions made can be controversial. Conservation areas can be created where areas are considered to be environmentally or historically important.

One of the most important responsibilities of the district council, and sometimes the county council, is to consider and process the applications for permission for new developments or the change of use of land or buildings. Applications range from house extensions to industrial estates and controversial issues like waste disposal, and all applications must conform to the local planning requirements. Without this control, there would be no way of making sure that local planning policies were fully implemented or that a building was not an eyesore. *Development control* is therefore concerned at a very detailed level with the use of land and the quality of the environment.

A day in the life of Carol – local plan officer

8.35   Met Emergency Planning Officer in the lift – discussed the availability of maps and who could produce them in emergencies.

8.40   Arrived in the office. Glanced at the post to check for anything urgent. Spoke to the Planning Technician about progress with modifications to the Local Plan, which will be published shortly.

9.00   Met with the Head of Development Control, Borough Engineer, Head of Property Services and the council's solicitor concerning agreements to be reached over a bid made by the county council for a local company.

9.50   Spoke separately to the Borough Engineer concerning the data required in applying for road scheme finance – which should be provided by the county council as Highway Authority.

9.55   Met my Principal Planning Officer to give urgent consideration to a Planning Assistant's contract of employment.

10.00   Attended a regular meeting with the Head of Development Control to discuss current planning issues.

11.15   Grabbed a cup of coffee before returning phone calls, principally one from English Partnership (EP) concerning valuation of a site within the Wet Dock which the council is considering purchasing, with EP's help.

12.00   Met with my Principal Planning Officer to discuss the aspects of the Inspector's Recommendations which our council is unable to accept. Agreed the list to be published within the Schedule of Proposed Modifications.

12.15 Spoke to council's solicitor about the above.

12.30 Met my Principal Planning Officer to continue discussion of the personnel matter.

1.00 Lunch time! Rushed into town. Ate a sandwich in the car.

2.00 Meeting with the Countryside Project Officer to decide agenda of a future meeting to further the town's Fringe Initiative in countryside management.

2.30 With the Head of Development Control, attended our weekly meeting with the Corporate Director.

3.40 Meeting with Head of Property Services concerning English Partnership, their potential purchase of the Wet Dock site and proposed development of student accommodation.

4.00 Meeting with representatives from Property Services, Legal Services and my Principal Planning Officer concerning a development brief for the new airport site. The Principal Planning Officer agreed to update and circulate the brief, requesting feedback from concerned bodies within a week – a saving in time for me.

Saw designer who is working on the Schedule of Modifications for publication in 10 days time – but only had time to wave!

5.00 Left early to take daughter home! Quite often, following up action points from the day's meetings and report writing can extend into the evening, meaning supper has to wait until 8.30–9.00pm. No evening meetings to attend in the town this week!

In summary, a fairly normal schedule, but with no external visits.

Who employs planners?

- Local government – councils and metropolitan authorities.
- Central government – the Civil Service, especially the Department of the Environment.
- Private companies such as developers and water companies.
- Planning and architectural consultancies.
- Larger estate agencies.
- Private planning consultancies.
- Organisations like the National Park authorities, urban development corporations, the Civic Trust and other community agencies.
- Foreign employers – British-trained planners can work overseas. Their qualifications are recognised within the European Union.

Local authority work

Most planners spend at least some of their career with a local authority. Local authorities are still the main employers, but increasingly planners are employed by others, as indicated above. When the services of planners are required, it is now compulsory for local authorities to ask for tenders (i.e. plans and costings for a job) from private planning consultants, architects, building companies, etc, not just from their own planning department. This also means that more work is done by planners working in private practice.

However, local authorities will certainly continue to be major employers. In a county council, a planner often works as part of a team, but is usually responsible for one particular aspect of the study – such as predicting future population levels, studying housing problems and needs, or investigating the need to protect attractive areas or how to use them for leisure purposes.

In a district or borough council, a planner usually works on local plans or development control. A local plan job often includes the detailed study of a particular area, its needs, its problems, and its potential. In development control, the work involves investigating the implications of peoples' applications for development – visiting each application site and investigating any particular problems, and then interpreting the local authority's policies in that particular situation.

Within planning departments, there are sometimes positions for professionals in other fields – such as people qualified in tourism, or specialists in industrial development.

EDUCATION AND TRAINING

Many of the subjects studied on a town planning degree – economics, survey methods, law, accounting, statistics and computing – have applications in other professions, and it would be possible to use a degree in planning as a basis for entry to other careers. To enter other related careers such as surveying, housing management and architecture or landscape architecture, it should be noted that the appropriate training would have to be undertaken.

To train as a professional **town planner** you could study for an accredited first degree (BA or BSc) plus diploma (DipTP) course in planning which satisfies the academic requirements of the Royal Town Planning Institute. The degree courses are called various names like *environmental planning, town and country planning, planning studies,* etc. A few universities offer planning degrees on a part-time basis.

Another route is to take a degree in a relevant subject such as geography, architecture, economics, etc, followed by a full-time postgraduate planning course, or study part-time whilst working. This route is a means of keeping other career options open.

Entry requirements
All of the above routes normally require A levels in two or three suitable subjects, such as geography, economics, mathematics, history, sciences, etc. None of these subjects is essential, but any could provide a useful background. An appropriate Advanced GNVQ is also acceptable, perhaps with additional units or an A level. GCSEs at grade C in a fairly wide range of subjects, to include mathematics and English, would also normally be required. Check the exact requirements of the course and institution of your choice: your careers library should have the appropriate higher education reference books. The Royal Town Planning Institute publishes a list of accredited courses.

Adults: note that maturity and previous experience may mean that stated entry requirements can be relaxed.

A study route which does not stipulate particular entry requirements is also available. This is a distance learning course provided by the Open University and a consortium of planning schools. This modular course satisfies the academic requirements of the Royal Town Planning Institute. This route is likely to take the average student eight years to complete, although a graduate entry route is available which reduces the course length.

Planning support staff
You need not completely abandon the idea of working in planning just because you are unlikely to get the high qualifications needed to become a qualified planner – there may be an opening for you as a **planning technician** or **planning administrator**. Technicians and administrators undertake the more routine practical and technical work in planning departments, and can also do some quite specialised work on computer systems, statistics and surveys. There are opportunities for career progression to **senior** or **principal technician**.

GETTING STARTED

The normal minimum requirements are four GCSEs at grade C, to include maths and a subject showing the use of English, or an appropriate Intermediate GNVQ, which would qualify you to study for BTEC National and then Higher National qualifications in Land Administration. Some employers prefer to recruit A level/Advanced GNVQ trainees. Sometimes, graduates may be recruited into this level.

If you started this way and achieved suitable BTEC qualifications by part-time study, or followed the distance learning programme mentioned above, you could still become professionally qualified in the long term.

Most town planning technicians and administrators start as school- or college-leavers, but there is no formal upper age limit to prevent mature entrants from starting in this career. Young people can enquire about training opportunities through their local careers service. Adults can ask at their local Jobcentre about possible government-sponsored training.

WASTES MANAGEMENT

We live in a world which produces remarkable quantities of waste – domestic rubbish, industrial effluents, radioactive by-products of nuclear power stations and so on. We only notice the work of wastes managers when something goes wrong, such as thousands of dead fish floating in a river or terrible smells from a factory. Controlling and disposing of these wastes is a very technical and complex process, and a high standard of education is required, often to degree level.

Over 100 million tonnes of waste are dealt with each year in this country. 20 million tonnes of it comes from our own homes, over half of which is kitchen waste and paper.

Waste collection

Household waste, and some commercial waste, is collected on a weekly basis. Local councils may run the service, but outside contractors are increasingly being used. Many large industrial companies dispose of their own waste.

Waste disposal

Most household waste is disposed of into landfill sites, where burying the waste allows it to decompose naturally. Several things have to be considered, such as selecting the site, building access roads, landscaping, the effect on the people living nearby, and protecting water sources. Certain industrial and chemical wastes have to be dealt with in different ways. They can be burnt at very high temperatures, or treated by special processes.

Waste recycling

Nowadays, there is much more interest in recycling wastes, and there are recycling sites all over the country where the public can bring paper, glass, aluminium cans, etc. Local authorities are being encouraged to set up recycling plants where, for example, waste can be burned to produce heat for domestic heating, and electricity for local use.

Wastes management, therefore, may involve knowing something about chemistry, civil engineering, geology and economics. There are other problems to be solved, too, such as the storage and transport of wastes. There are legal and economic considerations, with strict laws on pollution and the ways in which certain wastes can be disposed of.

So there's a lot to waste disposal! This does not mean that a wastes manager needs degrees in engineering, law, chemistry and geology. However, he or she would need to know where to go for specialist advice, and must be able to interpret the results of a study or a consultant's report.

EDUCATION AND TRAINING

Managers of sites which require a waste management licence are required by law to hold Certificates of Technical Competence. The route to achieving these is to gain the relevant NVQs which have been developed by the Waste Management Industry Training and Advisory Board (WAMITAB). The NVQs can be achieved while in employment. Previous study, qualifications and experience does not exempt people from the requirement, but can be assessed and accredited towards the NVQs.

The background of wastes managers is very varied. The educational standard required ranges from about five GCSEs at grade C to a degree. If you enter the industry prior to degree level, there are a variety of part-time and distance-learning courses you could take, leading to relevant qualifications. One example is a Higher National Certificate in Wastes Management, designed in conjunction with the Institute of Wastes Management, which is offered at some colleges, for which the minimum entry requirement is generally one A level or the equivalent. The NEBSM Certificate in Supervisory Management, with a Wastes Management option, is another relevant qualification, and the Open University offers a Diploma in Pollution Control.

If you wish to continue with full-time education to degree level before entering employment, there are a number of degree courses which include modules on waste management or environmental pollution control. Look out for degree courses with titles like *environmental risk management, wastes management and the environment*, or *environment and resource management*, for example. Graduates in subjects varying from civil engineering, environmental science, geology, hydrogeology, biology and chemistry to business management subjects can also enter waste management. The minimum entry requirement for degree courses is two A levels or the equivalent, and supporting GCSEs: check with individual institutions for specific requirements.

Postgraduate courses in waste management and environmental protection are available at a number of universities.

Professional training for the industry is overseen by the Institute of Wastes Management. Relevant courses, such as those mentioned above, can count towards full membership of the Institute. The Institute produces lists of relevant courses, including HNC, degree and postgraduate courses, and can advise on routes into the industry.

THE WATER INDUSTRY

When you drink water from your kitchen tap, or pull out the sink plug to let dirty water flow away, you probably don't think of where it's come from or where it goes. Yet over 50,000 people work in the water industry, trying to ensure that the supply is always there and safe to use. Research and engineering jobs are usually done by graduates, but there are also opportunities at craft and technician level for those with fewer qualifications.

The water cycle

The water in your tap has been collected in reservoirs, or may have come from a borehole deep underground. It is filtered, plumbed into mains and piped to your home. The dirty water flows away down sewers to the treatment works, where it emerges cleaned, purified and fit to start the cycle all over again.

The main organisations involved in water resources and supply are the water companies and the Environment Agency. Water supply and sewage disposal are managed by the water companies. River management, flood control, land drainage and pollution monitoring are the responsibility of the Environment Agency. Both the water companies and the Environment Agency are involved in fisheries and leisure and amenity management.

There are ten major water service companies in England and Wales. There are also 22 small companies supplying water only.

The Environment Agency has its headquarters in Bristol but also has eight regional centres. A very high proportion of its staff are involved in technical, scientific and engineering work.

There is a wide range of jobs in the water industry, from water bailiffs to geologists, lorry drivers, laboratory technicians, administrators and accountants. Some jobs are highly specialised, while others are really much the same as in many other commercial and public service organisations.

Jobs in the water industry

The various levels at which staff are employed are described below. Recruitment has been very low over the recent past, and it is unlikely to improve in the near future. You would need to contact employers to find out what the exact situation is in your area.

Operators and craft-level jobs

These involve laying and installing water mains and services, inspection of water fittings, operating pumping stations, operating treatment works, concrete construction, transport, clerical work and waterkeeping.

Technicians and supervisory jobs

These involve water supply and treatment, water resources, distribution, sewage treatment and disposal, laboratory work, civil engineering and design, mechanical engineering, finance and general administration.

Degree level

Posts are available for graduate trainees from a wide range of subject areas, including science, engineering and environmentally related subjects, as well as graduates from business, finance and information technology. Graduates are involved in management, research, engineering, financial control and personnel.

TRAINING

All jobs in the water industry involve some training, and the companies have well-organised training for new entrants. This can involve both part-time college courses and in-company training. The Certificate and Assessment Board for the Water Industry (CABWI) offers NVQs at levels 2 and 3 for water services operations and distribution, and NVQs from levels 1 to 3 in laboratory operations. Qualification requirements, entry levels and training offered vary from employer to employer, but a general outline is given below:

Operators – in water supply, sewage treatment and land drainage, reservoir attendant. Qualifications are not essential, although they are always helpful to offer. Training would be mainly on-the-job.

Craft and technician trainees – in mechanical, electrical and electronic engineering. Qualifications required may be four GCSEs at grade C or an Intermediate GNVQ for some positions, or appropriate A levels, BTEC National or Advanced GNVQ for higher-level technician posts. Training is generally on-the-job with day release, directed towards NVQ levels 2 or 3, or BTEC qualifications.

Graduates – in both the scientific and engineering field and in business services would follow a training programme lasting up to three years; the training would be based in the workplace with short specialist courses and possibly the opportunity to study for relevant professional qualifications or postgraduate qualifications.

HYDROGRAPHIC SURVEYING

Hydrographic surveyors produce charts of estuaries, har-bours, rivers, lakes and the seabed. High and low watermarks need to be established and the position of rocks, sandbanks and wrecks must be charted. Surveyors must discover whether channels used by shipping are being altered by silting or erosion, and at what rate.

Hydrographers may work for the Royal Naval Hydrographic Service, for port and river authorities, dredging companies, firms of consulting engineers, water companies and organisa-tions involved in land reclamation, fisheries work, offshore exploration activities, pipe and cable laying, building sea defences and marinas. The work usually involves periods on board survey vessels, drilling rigs, etc.

TRAINING

The University of Plymouth offers a BSc in hydrographic sur-veying; a one-year, postgraduate diploma course, leading to the DipHS qualification; and a one-year MSc in hydrography. Many hydrographic surveyors start by training in land survey-ing, as the basic principles involved are the same, and then take the postgraduate course. It is also possible to join the Royal Navy with a view to entering the RN Hydrographic Service.

Technical surveyor

Quite a lot of the practical surveying is carried out by technical surveyors. There are full-time and part-time BTEC National

and Higher National courses in land and hydrographic surveying at technician level. The Society of Surveying Technicians
can give further details.

OCEANOGRAPHY

The oceans cover 70% of the earth's surface, and we are all affected, to some extent, by their behaviour. Understanding how oceans work, and how we can best both use and conserve the resources they offer, is the work of oceanographers. Although there are some openings for support staff with fewer qualifications, oceanography is a graduate/postgraduate profession.

Oceanography is the result of a collaboration between many different science disciplines. Any one piece of research may involve contributions from several different specialists. The work is laboratory-based, but most oceanographers also carry out regular work at sea.

Marine physics involves the study of tides, waves, currents and ocean circulation.

Marine geology is concerned with the sea bed, its formation, composition and structure.

Marine chemists look at the chemical composition of sea water, marine organisms and sediments.

Marine biologists study the many life forms which exist in sea water, from limpets and barnacles to the mysterious creatures living in the complete darkness of the ocean depths.

Meteorologists research the effect of the oceans upon climate.

Oceanography has practical applications in areas such as the fisheries, mineral extraction and shipping management, but this does not mean that oceanographers are in demand in these industries. In the main, oceanographers work in universities or research institutes, and the results of their work are passed on to organisations who are interested. Sometimes a particular piece of research may be sponsored by an organisation and carried out by a university or institute research team.

EDUCATION AND TRAINING

Oceanography is very much a career for graduates. In general, not only a first degree but also a postgraduate qualification such as an MSc or PhD will be needed. A range of first degrees is acceptable: chemistry, physics, biology, geology, geography and geophysics are the most usual, while mathematics is less common.

There are some first degree courses in which there are opportunities to study oceanography in combination with other subjects, such as geology. There are also degrees with titles like *earth sciences* or *environmental sciences*, which may include some oceanography. You'll need to check carefully in individual prospectuses and the higher education handbooks. Think carefully before you choose your course – unless you are determined on a career in oceanography, taking a vocational degree may limit career openings. You'll need good science and mathematics A levels or relevant Advanced GNVQs or BTEC National qualifications to be accepted on such degree courses, although there are some conversion courses available for people with arts A levels who wish to study science or technology at degree level.

Below graduate level, there are occasional opportunities for technical and support workers at the Institute for Oceanographic Sciences. Applicants are likely to need four

GCSEs at grade C, including English language, mathematics and a science, and preferably two A levels, or possibly Intermediate and Advanced GNVQs.

Adults: note that maturity and previous experience may mean that stated entry requirements can be relaxed.

Where are oceanographers employed?

In the United Kingdom, most employment is in research institutes financed by government funds, or in research teams in universities. Only very small numbers of graduates go into this work each year, and some posts may only be contracts for a fixed period. Oceanographers are also employed by some oil companies, the water industry and by marine survey and consulting companies. Work pressures can be high and job security lower than in research, but the financial reward can be good.

THE OFFSHORE OIL & GAS INDUSTRY

The offshore oil and gas industry is thriving, and demand is on the increase. New techniques for drilling faster and deeper are being developed. Staff work either at land-based offices or factories, or on offshore platforms all over the world – in Britain, mainly off the east coast of England and Scotland. Opportunities vary from jobs for graduate engineers and scientists to semi-skilled labourers or roustabouts.

Oil and gas are among the basic raw materials for modern industry. Over a quarter of our needs are now supplied from offshore installations. There are three main areas of work:

Exploration – geological and seismic surveys and their analysis; test drilling from drilling rigs. To find natural reservoirs of oil and gas which are economically worthwhile exploiting takes the combined efforts of a highly skilled and experienced team of geologists, engineers and technical staff.

Field development – deciding how to get oil out of a reservoir, setting up production facilities, drilling production wells. Operation engineers and technicians set up the production site.

Production and maintenance – operating and maintaining equipment. There are jobs on support vessels, on rig and platform construction and at terminals. The storing and refining of supplies of crude oil at terminals requires very careful planning and scheduling of operations.

What is the work like?

- very tough and demanding;
- dirty, wet and noisy;
- alcohol and smoking are banned;
- pay is good, with free travel;
- work offshore is on a shift basis, the standard pattern being 12 hours on, 12 hours off for two weeks, then two weeks ashore. Home life is very disrupted.

Workers usually have to undergo a survival course at their own expense before they can be offered a job. Once employed, they receive further safety and firefighting training.

GETTING STARTED

A minimum of three GCSEs at grade C, or the equivalent, are required. GCSEs must include English and maths. The minimum age for working on a rig is 18, but, in practice, most workers are aged between 20 and 35. Finding a job is not easy, though there is a high turnover of unskilled workers (many people can't put up with the conditions for long). Relevant experience or training helps. Recruitment is largely done by contractors in towns like Aberdeen or Lowestoft, except for graduates who are recruited directly by the oil companies and other employers. Workers can enrol for training courses while onshore, which may help their promotion prospects.

Oil rigs are not all 'men only' establishments. There are women technicians, mechanics, geologists, engineers, chefs, etc.

The UK offshore industry comprises internationally operating companies like Shell, BP, British Gas, Conoco; engineering contractors and service companies who carry out the specialist jobs; drilling and maintenance contractors, diving firms, mud analysts, cement-casing firms and suppliers of all the items the operation requires; design engineers; and the Petroleum Engineering Division of the Department of Trade and Industry.

Some of the jobs

Roustabouts do the basic labouring job on the platform, loading and unloading supply ships and helicopters, moving items to storage and general maintenance work. Strength and fitness are necessary, preferably with some knowledge of an engineering craft.

Roughnecks carry out the manual work of the drilling operation under the direction of the driller and derrickmen. The job is hard physical labour.

Derrickmen work high up on the derrick (or steel tower) handling the sections of the drill pipe under the driller's direction.

Drillers/assistant drillers perform the key role on the drilling platform. The driller operates the drilling equipment and directs the work of the drilling crew. With the right abilities and temperament it could take three to five years to get from roughneck to driller. Some companies recruit assistant drillers from applicants with an engineering background.

Toolpushers oversee operations and may run the whole drilling rig, making sure everything goes smoothly and that necessary materials and equipment are available. Assistants to toolpushers are often graduate trainees getting experience of the work.

Mud loggers are highly trained geologists. The operating company must know what rock formations are being drilled through and when there are any signs of the precious hydrocarbons.

Production operators work under supervision on various aspects of production, storage and distribution.

Divers have to have undergone rigorous training in diving techniques; it can be a hazardous occupation. Most divers pay for their own training at a recognised diving school, and recoup the cost through the comparatively high rates of pay they then receive. Just being able to dive is not enough – there are usually practical tasks to be done underwater, so some engineering or construction skills are needed in addition.

Other jobs
Rigs also need welders, electricians, mechanics, medical workers and cooks, all of whom will have learned their skills and gained experience onshore. There are opportunities for administrative and clerical staff with IT skills, but the majority of these posts are shore-based.

Graduate jobs
Most graduates are qualified in chemistry, physics, engineering, maths, geology or geophysics. Some firms offer sponsorship to undergraduates.

Geologists/geophysicists carry out surveys and analyse figures to assess the prospects of finding oil or gas. They undertake research and advisory work, preparing reports and charts for the contracting company.

Petroleum engineers apply principles of mathematics, physics, chemistry and engineering to the economic recovery and processing of hydrocarbons.

Drilling engineers analyse drilling performance and factors affecting cost and efficiency.

Reservoir engineers estimate probable field production.

Subsea engineers design, maintain and operate the engineering systems and equipment under the sea.

GEOLOGY & GEOPHYSICS

Geoscientists produce geological maps and databases, explore for mineral resources and carry out ground surveys associated with a wide range of engineering and environmental projects. Geoscience involves the different science disciplines of geology and geophysics, and includes geochemistry. Most careers in this area require a good science-based degree, but there are also some opportunities for technicians with lower qualifications.

Much of the **geologist's** work is carried out in the 'field', by examining and sampling rocks in natural exposures, by drilling boreholes, or by carrying out geophysical surveys such as seismic surveys to 'sense' the subsurface geology. Specimens, fossils and minerals are collected and subsequently analysed in the laboratory, or processed by computer.

Exploration **geophysics** involves taking measurements with sensitive survey equipment on the surface or in boreholes to investigate the physical properties of concealed rocks. Exploration **geochemistry** involves taking rock, sediment or stream samples in the search for elements or compounds of possible economic importance. Both methods are widely used in the search for mineral resources.

Geoscientists present their work in the form of papers or technical reports, including geological maps and sections, and associated data.

Would I like it?

- You need to be in good health as you could be working in any country and climate, and under difficult physical conditions.
- You need accurate observation and objective judgement.
- Are you self-reliant?
- Qualities of leadership may be needed to organise and supervise the work of a large team.

EDUCATION AND TRAINING

Technician level

There are some openings in geology at technician level – mainly for laboratory-based posts, providing support to professional staff. Good GCSEs at grade C and preferably A levels (including sciences) or an appropriate Advanced GNVQ or BTEC National qualification are likely to be needed. There are relevant BTEC Higher National Diplomas. Usual entry requirements are one science A level pass, with another studied to A level, an Advanced GNVQ or a BTEC National Certificate or Diploma, plus four or five GCSEs at grade C.

Professional level

To work professionally in geology or geophysics, you normally need an honours degree. At present, about half the geologists entering first jobs have higher degrees (MSc or PhD). Most universities and some other higher education institutions offer geology courses, but a smaller number offer geophysics. You can study geology with a subsidiary subject such as chemistry, physics, mathematics or biology. There are also combined honours courses where geology is one of two main subjects. A few courses are offered in specialist aspects of geology, such as engineering geology. Entry requirements vary; for many geology courses at least two A levels in sciences and mathematics are required, the most useful subjects being maths, physics, chemistry and biology. Geography or geology at A level are acceptable but not essential. Relevant Advanced GNVQs or BTEC National qualifications may also be acceptable. Check with individual institutions.

For geophysics degrees, mathematics and physics A level are normally needed – or possibly the equivalent qualifications mentioned above. Postgraduate study or training may also be necessary, perhaps in a specialisation such as hydrogeology. The choice of postgraduate course (from approximately 70 in the UK) will depend on the subsidiary subjects studied at first degree level. Graduates of other disciplines may also specialise in branches of geology. A physicist may become a geophysicist, a zoologist can become a palaeontologist, and a chemist may take a specialised course in geochemistry.

Where do geologists find work?

There are about 6000 geoscientists employed in the UK. Most vacancies are in the oil industry, and many posts, even with British-based firms, entail periods of service overseas of up to five years. The civil engineering industry employs a small but increasing number of geologists for site foundation work, dams and road building. Opportunities are also growing in environmental and waste-disposal industries.

To get an idea of salaries and vacancies, read adverts in the daily and Sunday papers (*Independent*, *Telegraph*, *Times*, *Guardian*) and magazines such as *New Scientist*.

In the UK many geologists, geophysicists, geochemists, engineering geologists and hydrogeologists are engaged in the British Geological Survey, where staff may be required to serve overseas for a part of their career. The Survey is administered through the Natural Environment Research Council. A few posts are also available in other government and associated research establishments, such as the Department of the Environment, the Building Research Station, the Soil Survey, museums, the universities, to a limited extent in schools, and in some industrial concerns such as civil engineering contractors, oil companies, the metals industry and geological/geophysical consultants.

Overseas, there are also posts with oil companies, mining groups and industrial firms. The Overseas Development Administration recruits experienced geologists for contract appointments in many Commonwealth countries and dependencies. There are very occasional vacancies in the British Antarctic Survey.

Geoscience graduates can qualify as teachers through a one-year postgraduate course, as long as their degree course contained enough of the national curriculum subjects like chemistry, physics or geography.

About 3500 **technical staff** are employed in the geosciences, about half of whom work in the oil industry; others are found in civil engineering firms or research institutions. The jobs are very diverse, and include working out on site and in the laboratory, assisting with mapping, the analysis of samples or the preparation of data.

FOR FURTHER INFORMATION

CAREERS USING GEOGRAPHY

Geographical Association – 343 Fulwood Road, Sheffield S10 3BP. Tel: 0114 267 0666.

Royal Geographical Society – 1 Kensington Gore, London SW7 2AR. Tel: 0171 589 5466.

CARING FOR THE ENVIRONMENT

British Trust for Conservation Volunteers – 36 St Mary's Street, Wallingford, Oxfordshire, OX10 0EU. Tel: 01491 839766.

COSQUEC (Council for Occupational Standards and Qualifications in Environmental Conservation) – Executive Coordinator, The Red House, Pillows Green, Staunton, Gloucester GL19 3NU. Tel: 01452 840825.

Countryside Commission – John Dower House, Crescent Place, Cheltenham GL50 3RA. Tel: 01242 521381.

Countryside Management Association – Administration Department, Drury Lane, Knutsford, Cheshire WA16 6HB. Tel: 01565 633603. Produces *Countryside Jobs Advice Network*, a weekly listing of countryside jobs of all kinds.

English Nature – Enquiry Service, Room IW, Northminster House, Peterborough, Cambridgeshire PE1 1UA. Tel: 01733 455100.

Environment Agency – Rio House, Waterside Drive, Aztec West, Almondsbury, Bristol BS12 4UD. Tel: 01454 624400.

Environment Council – 21 Elizabeth Street, London SW1W 9RP. Tel: 0171 824 8411.

Institution of Environmental Sciences – 14 Princes Gate, Hyde Park, London SW7 1PU. Tel: 01778 394846.

National Trust – see *Yellow Pages* for the address and telephone number of your regional office. Their Careership Office is at Lanhydrock, Bodmin, Cornwall PL30 4DE. Tel: 01208 74281.

Natural Environment Research Council (NERC) – Polaris House, North Star Avenue, Swindon, Wilts SN2 1EU. Tel: 01793 411500.

See CRAC/Hobsons Press *Degree Course Guide: Geological and Environmental Sciences*. There are also Guides covering other subjects such as Agricultural Sciences, Landscape Architecture (under Architecture), Biological Sciences, etc. HND courses, degrees and postgraduate qualifications are listed in the *Directory of Environmental Courses*, published by the Environment Council: send an A4 self-addressed, stamped envelope (79p postage).

Careers in Environmental Conservation, published by Kogan Page.
Working in the Environment, published by COIC.
Working in Geography, published by COIC.
Environmental Careers Handbook, published by Trotman.
Working with the Environment, published by Vacation Work.
Environment Post, published monthly by ADC Environment, 33 Nobel Square Basildon, Essex SS13 1LT. Tel: 01268 724259. The magazine has an appoinments section.

Meteorology

Institute of Oceanographic Sciences – Deacon Laboratory, Brook Road, Wormley, Godalming, Surrey GU8 5UB. Tel: 01428 684141. Please send a large stamped, addressed envelope (57p postage) with requests for information.

Meteorological Office – Personnel Department (Recruitment), Room 615, London Road, Bracknell, Berkshire RG12 2SZ. Tel: 01344 856032.

Royal Meteorological Society – 104 Oxford Road, Reading, Berkshire RG1 7LL. Tel: 0118 956 8500. Send a stamped, addressed envelope for careers information.

Cartography

British Cartographic Society – Geology & Cartography

Division, School of Construction and Earth Sciences, Oxford Brookes University, Gipsy Lane, Headington, Oxford OX3 0BP. The Society produces a booklet, *Careers in Cartography*, which gives details of courses and lists possible employers: it is available from R.W. Anson, telephone 01865 483346.

SURVEYING

Architects and Surveyors Institute – 15 St Mary Street, Chippenham, Wiltshire SN15 3WD. Tel: 01249 444505.

Association of Building Engineers – Jubilee House, Billing Brook Road, Weston Favell, Northampton NN3 8NW. Tel: 01604 404121.

Association of Consulting Engineers – Alliance House, 12 Caxton Street, London SW1H 0QL. Tel: 0171 222 6557.

Chartered Institute of Building – Englemere, Kings Ride, Ascot, Berks SL5 8BJ. Tel: 01344 23355.

Chartered Surveyors' Training Trust – 9 Bentinck Street, London W1M 5RP. Tel: 0171 224 0205.

Civil Engineering Careers Service – 1 Great George Street, London SW1P 3AA. Tel: 0171 222 7722.

Construction Industry Training Board – Bircham Newton, King's Lynn, Norfolk PE31 6RH. Tel: 01553 776677.

Incorporated Society of Valuers and Auctioneers (ISVA) – 3 Cadogan Gate, London SW1X 0AS. Tel: 0171 235 2282.

Institution of Civil Engineering Surveyors – Newholme, 26 Market Street, Altrincham, Cheshire WA14 1PF. Tel: 0161 928 8074.

Institution of Structural Engineers – 11 Upper Belgrave Street, London SW1X 8BH. Tel: 0171 235 4535.

Royal Institution of Chartered Surveyors – Surveyor Court, Westwood Way, Coventry CV4 8JE. Tel: 0171 222 7000 or 01203 694757. Provides careers information on request.

Society of Surveying Technicians – Surveyor Court, Westwood Way, Coventry CV4 8JE. Tel: 0171 222 7000 or 01203 694757.

The *Ivanhoe Guide to Chartered Surveyors*, published by CMI Ltd, is available from bookshops, or may be held in your local careers library.

Building a Career has a section on building surveying work, and is available from the Construction Careers Service, CITB, Bircham Newton, King's Lynn, Norfolk PE31 6RH. Tel: 01553 776677.

Surveying, a careers booklet for graduates, available from CSU, Armstrong House, Oxford Road, Manchester M1 7ED. Tel: 0161 236 9816, ext 250/251.

MINING & QUARRYING

British Aggregate Construction Materials Industries – 156 Buckingham Palace Road, London SW1W 9TR. Tel: 0171 730 8194. Produce the booklets *Finding Out about Quarrying* and *Rocks Around You.*

Celtic Energy Ltd – Heol Ty Aberaman, Aberaman, Aberdare, Mid Glamorgan CF44 6LX. Tel: 01685 874 201.

Doncaster College – High Melton, Doncaster, South Yorkshire DN5 7SZ. Tel: 01302 553922.

Institute of Quarrying – 7 Regent Street, Nottingham NG1 5BS. Tel: 0115 9411 315. Contact the Press Officer if a group of interested students would like to visit a working quarry.

Institution of Mining and Metallurgy – Contact Peter Martindale, 44 Portland Place, London W1N 4BR. Tel: 0171 580 3802.

Institution of Mining Engineers – Dr Woodrow (Secretary), Danum House, 6A South Parade, Doncaster DN1 2DY. Tel: 01302 320486.

Mineral Industries Educational Trust – 6 St James's Square, London SW1Y 4LD. Tel: 0171 753 2117. Provides information on bursaries.

Mining Association of the UK – 6 St James's Square, London SW1Y 4LD. Tel: 0171 930 2399. Provides information on employers.

Quarry Products Training Council – Sterling House, 20 Station Road, Gerrards Cross, Bucks SL9 8HT. Tel: 01753 891808.

RJB Mining plc – Harworth Park, Blyth Road, Harworth, Doncaster, South Yorkshire DN11 8DB. Tel: 01302 751751.

Exploration, Extraction and Protection of Natural Resources, a careers booklet for graduates, available from CSU, Armstrong House, Oxford Road, Manchester M1 7ED. Tel: 0161 236 9816, ext 250/251.

Estate Agency

National Association of Estate Agents – Arbon House, 21 Jury St, Warwick CV34 4EH. Tel: 01926 496800.

College of Estate Management – Whiteknights, Reading RG6 2AW. Tel: 01734 861101.

Residential Estate Agency Training and Education Association (REATEA) – The Avenue, Brampford Speke, Exeter EX5 5DW. Tel: 01392 841194.

Society of Surveying Technicians – Surveyor Court, Westwood Way, Coventry CV4 8JE. Tel: 0171 222 7000 or 01203 694757.

Valuation & Auctioneering

Association of Building Engineers – Billing Brook Road, Weston Favell, Northampton NN3 4NW. Tel: 01604 404121.

Incorporated Society of Valuers and Auctioneers (ISVA) – 3 Cadogan Gate, London SW1X 0AS. Tel: 0171 235 2282.

Institute of Revenues, Rating and Valuation – 41 Doughty Street, London WC1N 2LF. Tel: 0171 831 3505.

Landscape Architecture

British Association of Landscape Industries – Landscape House, 9 Henry Street, Keighley, West Yorkshire BD21 3DR. Tel: 01535 606139.

Landscape Institute – 6–7 Barnard Mews, London SW11 1QU. Tel: 0171 738 9166.

Professional Careers in Landscape Architecture, Sciences and Management – available from The Landscape Institute (address above). The Institute has a 35-minute video entitled *For People and the Planet*

– careers and the landscape professions, which can be purchased or is available to hire.

Town & Country Planning
Royal Town Planning Institute – 26 Portland Place, London W1N 4BE. Tel: 0171 636 9107.
Scottish Office Development Department – Victoria Quay, Edinburgh EH6 6QQ. Tel: 0131 556 8400.
Society of Town Planning Technicians – c/o 26 Portland Place, London W1N 4BE (correspondence only).

Architecture, Landscape Architecture, and Town & Regional Planning – an AGCAS Graduate Careers Information Booklet, available from CSU, Armstrong House, Oxford Road, Manchester M1 7ED. Tel: 0161 236 9816, ext 250/251.

Wastes Management
Environmental Services Association – 154 Buckingham Palace Road, London SW1W 9TR. Tel: 0171 824 8882.
Institute of Wastes Management – 9 Saxon Court, St Peter's Gardens, Northampton NN1 1SX. Tel: 01604 20426.
National Association of Waste Disposal Contractors – Mountbarrow House, 6–20 Elizabeth Street, London SW1W 9RB. Tel: 0171 824 8882.
Waste Management Industry Training and Advisory Board (WAMITAB) – PO Box 176, Northampton NN1 1SB. Tel: 01604 321950.

The Water Industry
Board for Education and Training in the Water Industry (BETWI) – 1 Queen Anne's Gate, London SW1H 9BT. Tel: 0171 957 4524.
Certificate and Assessment Board for the Water Industry (CABWI) – 1 Queen Anne's Gate, London SW1H 9BT. Tel: 0171 957 4523.
Environment Agency – Rio House, Waterside Drive, Aztec West, Almondsbury, Bristol BS12 4UD. Tel: 01454 624400.

Environment and Heritage Service – Environment Protection, Culvert House, 23 Castle Place, Belfast BT1 1FY. Tel: 01232 254754.

Institution of Water Officers – Heriot House, 12 Summerhill Terrace, Newcastle upon Tyne NE4 6EB. Tel: 0191 230 5150. Deals with enquiries relevant to membership, but does not provide careers information.

Scottish Environmental Protection Agency – Erskine Court, The Castle Business Park, Stirling FK9 4TR. Tel: 01786 457700.

Water Companies Association – 1 Queen Anne's Gate, London SW1H 9BT. Tel: 0171 222 0644.

Water in Scotland – Scottish Office, Environment Department, 27 Perth Street, Edinburgh EH3 5RB. Tel: 0131 556 8400.

Water Services Association – 1 Queen Anne's Gate, London SW1H 9BT. Tel: 0171 957 4567.

Anglian Water – Ambury Road, Huntingdon PE18 6NZ. Tel: 01480 443000.

Dwr Cymru – Plas-y-Ffynnon, Cambrian Way, Brecon, Powys LD3 7HP. Tel: 01874 623181.

East of Scotland Water Authority – Head Office, Pentland Gait, Calder Road, Edinburgh EH11 4HJ. Tel 0131 445 4141.

North of Scotland Water Authority – Caledonia House, 63 Academy Street, Inverness IV1 1LU. Tel: 01463 245 400.

North West Water – Dawson House, Great Sankey, Warrington, Cheshire WA5 3LW. Tel: 01925 234000.

Northern Ireland Water Services – Northland House, 3 Frederick Street, Belfast BT1 2NS. Tel: 01232 244711.

Northumbrian Water – Abbey Road, Pity Me, Durham DH1 5FJ. Tel: 0191 383 2222.

Severn Trent Water – 2297 Coventry Road, Birmingham B26 3PU. Tel: 0121 722 4000.

South West Water – Peninsula House, Rydon Lane, Exeter EX2 7HR. Tel: 01392 446688.

Southern Water – Southern House, Yeoman Road, Worthing BN13 3NX. Tel: 01903 264444.

Thames Water – Nugent House, Vastern Road, Reading, RG1 8DB. Tel: 01734 591159.

Wessex Water – Wessex House, Passage Street, Bristol BS2 0JQ. Tel: 0117 929 0611.

West of Scotland Water Authority – 419 Balmoral Road, Glasgow G22 6NU. Tel: 0141 355 5333.

Yorkshire Water – West Riding House, 67 Albion Street, Leeds LS1 5AA. Tel: 0113 244 8201.

HYDROGRAPHIC SURVEYING

The Hydrographic Society – University of East London, Longbridge Road, Dagenham RM8 2AS. Tel: 0181 597 1946.

OCEANOGRAPHY

Southampton Oceanography Centre – Empress Dock, Southampton SO14 3ZH. Tel: 01703 595000. The Research Vessel Base can be found at this address.

Society for Underwater Technology – 76 Mark Lane, London EC3R 7JN. Tel: 0171 481 0750 – produces *Oceans of Opportunity*, an extensive careers pack with information on degree and postgraduate courses in oceanography and related subjects. A careers video is also available.

THE OFFSHORE OIL & GAS INDUSTRY

Institute of Petroleum – 61 New Cavendish Street, London W1M 8AR. Tel: 0171 467 7100.

Society for Underwater Technology – 76 Mark Lane, London EC3R 7JN. Tel: 0171 481 0750.

GEOLOGY AND GEOPHYSICS

British Geological Survey – Kingsley Dunham Centre, Keyworth, Nottingham NG12 5GG. Tel: 0115 936 3100 – send an A5 stamped, self-addressed envelope for a leaflet *Careers in the Geological Sciences*.

Geological Society – Burlington House, Piccadilly, London W1V 0JU. Tel: 0171 434 9944. Produces leaflets on careers in geology: send an A5 stamped, self-addressed envelope.

Exploration, Extraction and Protection of Natural Resources, from CSU, Armstrong House, Oxford Road, Manchester M1 7ED. Tel: 0161 236 9816, ext 250/251.

CRAC *Degree Course Guide: Geography and Geological Sciences*, at your local careers library.